BIOLOGICAL EFFECTS OF ELECTRIC AND MAGNETIC FIELDS OF EXTREMELY LOW FREQUENCY

BIOLOGICAL EFFECTS OF ELECTRIC AND MAGNETIC FIELDS OF EXTREMELY LOW FREQUENCY

Asher R. Sheppard, Ph.D.
Merril Eisenbud, Sc.D.

Institute of Environmental Medicine
New York University Medical Center

New York • New York University Press • 1977

Library of Congress Catalog Card Number: 76-39695
ISBN: 0-8147-2562-7

We are grateful to the following publishers for permission to utilize tables and figures which appeared in their publications:

Academic Press, Inc. for Fig. 1 and Table II by H. P. Schwan in *Adv. in Biol. and Med. Phys. 5:* 147 (1957).

Electric Power Research Institute for Figs. 8.2.1, 8.2.2, and 8.3.2, and Table 8.5.1 in "Transmission Line Reference Book 345 kV and Above, 1975" (1975).

The Electrochemical Society, Inc. for table by A. A. Marino in *J. Electrochem. Soc. 123* (1976).

The Institute of Electrical and Electronics Engineers, Inc. for Fig. 1 by T. D. Bracken in *IEEE Trans. on Power Apparatus & Systems PAS-95* (1976); and for Table III and Figs. 3, 4 and 5 by H. C. Barnes in *IEEE Trans. on Power Apparatus & Systems PAS-86* (1967).

The New York Academy of Sciences for Tables 1-4 by A. A. Marino in *Ann. N.Y. Acad. Sci. 238:* 436 (1974); and Fig. 1 and Tables 1 and 2 by S. Bawin in *Ann. N.Y. Acad. Sci. 247:* 74 (1975).

Plenum Publishing Corporation for Fig. 2 (p. 14) by H. L. Konig in "ELF and VLF Electromagnetic Field Effects," M. Persinger, ed. (1974).

Manufactured in the United States of America

ACKNOWLEDGMENTS

This study would not have been possible without the post doctoral fellowship granted to Dr. Sheppard by the National Institute of Environmental Health Sciences and the NIEHS Center Program grant, which provided many forms of institutional support. We are also grateful for assistance provided by the American Electric Power Service Corporation.

Michael Marmor of this Institute was particularly helpful as a critical reviewer of the manuscript, and we acknowledge in addition the helpful suggestions of others who read the manuscript, including Thomas S. Ely, Sol Michaelson, W. Ross Adey, Brendan Ware, and Dan Shih. We are also indebted to James Rabinowitz and George Pack for many hours of stimulating and productive discussion.

Christine Singleton and Norene Block were very helpful in the difficult task of locating many of the references, and Gordon Cook and Roger Sparling provided many of the charts and line drawings.

Finally, in addition to acknowledging the assistance of many others too numerous to thank individually, we wish to express our appreciation to Eleanor Clemm, who typed the final manuscript and provided dedicated assistance in an editorial and coordinating capacity.

CONTENTS

Foreword xv

Chapter 1—INTRODUCTION

1.1 Note on the Units Used in this Report 1-4
1.2 Note on Books, Review Articles and Bibliographies Available in the Literature 1-4
1.3 Note on References Following Chapters 1-7

Chapter 2—SUMMARY

2.1 Fields in the Environment 2-1
2.2 Fields in the Body 2-1
2.3 Fields Involved in Normal Biological Functions 2-7
2.4 Field Effects in Animals and Humans 2-9
- Neurophysiological and Behavioral Effects
- Biochemical Effects
- Clinical, Physiological and Behavioral Effects
- Effects on Reproduction

2.5 Effects of Strong Magnetic Fields 2-23
2.6 Conclusions and Discussion 2-25
2.7 References 2-31

Chapter 3—ENVIRONMENTAL EXPOSURES

3.1 References 3-8

Chapter 4—BIOPHYSICAL PRINCIPLES AND A BIOPHYSICAL DESCRIPTION OF EXPOSURE TO ELF FIELDS

4.1 The Quasi-Static Approximation 4-1
4.2 Energy Density in the Electric Field 4-1
4.3 Dielectric Constant 4-2
4.4 Conductivity 4-4
4.5 Models of a Conductor in an Electric Field 4-4

4.6 Electric Field Distortion 4-9
4.7 Fields Inside the Body 4-14
4.8 Power Loss Inside the Body 4-14
4.9 Summary of Conducting Spheroid Model 4-15
4.10 Current Flow into Cells 4-15
4.11 Cell Size, Shape, and the Membrane Potential in an Electric Field 4-16
4.12 Summary of Electrical Description of a Body in an AC Electric Field 4-17
4.13 Electrostatic Induction 4-17
4.14 Human Capacitance 4-18
4.15 Discharge Effects 4-18
4.16 Electrophoresis and Dielectrophoresis 4-19
4.17 Electric Induction by AC Magnetic Fields 4-20
4.18 Electromagnetophoresis 4-20
4.19 Problems in Dosimetry 4-21
4.20 Electric Shock Currents 4-23
4.21 References 4-26

Chapter 5—HEMATOLOGICAL AND BIOCHEMICAL EFFECTS OF ELF FIELDS

5.1 The Biochemical Time Scale 5-1
5.2 Electric and Magnetic Field Effects on Blood Composition and Chemistry 5-2
- Asanova and Rakov, 1966
- Kouwenhoven *et. al.,* 1967; Singewald et al., 1972
- Hauf, 1974
- Blanchi *et al.,* 1973a
- Beischer *et al.,* 1973
- de Lorge, 1974
- Krumpe and Tockman, 1972; Krumpe and Tockman, 1974
- Comparative Summary of the Foregoing Reports

5.3 Effects on Blood Proteins and Steroid Hormones 5-14
- Marino *et al.,* 1974a; 1974b; 1976a; 1976b; 1976c
- Friedman and Carey, 1972

5.4 Effects on Microsomal Metabolism 5-27
- Riesen *et al.,* 1971; Straub and Carver, 1975

5.5 Effects on Respiration 5-28
- Greenberg and Ash, 1974; Greenberg, 1975

5.6 Calcium and GABA Efflux in Brain Tissues 5-29
- Bawin and Adey, 1975; Kaczmarek and Adey, 1974
- Bawin *et al.,* 1975

5.7 Effects on Protein Synthesis 5-33
- Harrington and Becker, 1973
- Bassett and Hermann, 1968

5.8 Effects on Immunological Ability 5-34
- Fischer, 1973

5.9 Effects on ATP Metabolism 5-34
- Kolodub and Yevtushenko, 1974; 1973

5.10 Effects on Bone Growth and Fracture Repair 5-35
- Bone Growth in Electric Fields
- Bone Growth Using a Pulsed Magnetic Field
- Electrical Influence on the Mitotic Rate in Bone
- Polarity Effect
- Microscopic Evidence

5.11 Summary 5-38

5.12 References 5-40

Chapter 6—PHYSIOLOGICAL AND BEHAVIORAL EFFECTS IN HUMANS

6.1 Perception of Low Level Electric and Magnetic Fields. 6-1
- Hamer, 1968
- Konig, 1974b
- Friedman *et al.,* 1967
- Milburn, 1971
- Wieske, 1963
- Schmitt and Tucker, 1973

6.2 Perception, Behavioral and Physiological Responses in a Magnetic Field. 6-5
- Beischer *et al.,* 1973

6.3 Psychomotor Function and Cognitive Skills in a Magnetic Field 6-5
- Gibson and Moroney, 1974

6.4 Effect of "Zero" Magnetic Field 6-6
- Beischer, 1965

6.5 Physiological Effects on Substation Workers 6-7
- Sazonova, 1967
- Asanova and Rakov, 1966
- Belyaeva and Stroikova, 1975

6.6 Effects of Electric Discharges (Electrostatic Induction) in Man 6-13
- Krivova, 1968
- Krivova *et al.,* 1975a

6.7 Physiological Studies on Discharge Effects in Man and Animals 6-14
- Takagi and Muto, 1971

6.8 Physiological Studies in Man. 6-15
- Hauf, 1974
- Kouwenhoven *et al.,* 1967; Singewald et al., 1972
- Johansson *et al.,* 1973
- Krumpe and Tockman, 1972; 1975

6.9 Effects on Biological (Circadian) Rhythms 6-17
- Wever, 1967;1968; 1974

6.10 Correlation of Human Behavior and Geomagnetic Storms 6-18
- Friedman and Becker, 1963

6.11 Miscellaneous Reports 6-18
- Tromp, 1949

6.12 Dowsing: a Magnetic Perception? 6-20
- Tromp, 1968; 1949

6.13 References 6-21

Chapter 7—PHYSIOLOGICAL AND BEHAVIORAL EFFECTS IN NON-HUMAN SPECIES

7.1 Central Nervous System Effects Produced in Monkeys by Electric Fields 7-1
- Gavalas-Medici and Magdaleno, 1975
- Gavalas *et al.,* 1970

7.2 Central Nervous System Effects in Cats Using an ELF Modulated Radio Frequency Carrier Wave 7-10
- Bawin *et al.,* 1973

7.3 Electric Fields in Brain Tissue 7-11

7.4 Behavioral and Physiological Studies of Monkeys in Magnetic and Electric Fields. 7-12
- Grissett and deLorge, 1971; deLorge, 1974
- deLorge, 1972; 1973a; 1973b

7.5 EEG and ECG Changes in a 50 Hz Electric Field 7-15
- Blanchi *et al.,* 1973b

7.6 EEG Changes in an Electrostatic Field 7-15
- Lott and McCain, 1973

7.7 EEG Changes in Monkeys in High Magnetic Fields 7-17
- Beischer and Knepton, 1966

7.8 Electric Field Effects on Neurons 7-17
- Gengerelli and Holter, 1941
- Terzuolo and Bullock, 1956

7.9 Mitotic Cycle in Slime Mold 7-18
- Marron *et al.,* 1975; Goodman *et al.,* 1975

7.10 Fertility and Pathology in Male Mice................. 7-19
- Knickerbocker *et al.,* 1967

7.11 Activity in Mice....................................... 7-20
- Moos, 1964

7.12 Effects on Mice....................................... 7-20
- Belyaeva and Stroikova, 1975

7.13 Psychological Behavior................................ 7-21
- Persinger and Pear, 1972; Persinger *et al.,* 1973
- Persinger, 1974

7.14 Electric and Magnetic Field Effects in Animal Migration, Navigation and Communication 7-22
- Rommel and McCleave, 1972; 1973a; 1973b
- Bullock, 1973; Hopkins, 1974
- Southern, 1973; 1975
- Walcott, 1974
- Blakemore, 1975
- Graue, 1974

7.15 Electric Field Effects on Marine Animals 7-27
- Straub *et al.,* 1972

7.16 Effects on Conditioned Behavior in Pigeons and Rats 7-28
- Marr *et al.,* 1972

7.17 Effects on Growth and Development.................... 7-28
- Krueger *et al.,* 1972
- Durfee *et al.,* 1975a
- Smith, 1975
- Yates *et al.,* 1975
- Durfee *et al.,* 1975b

7.18 Electrical Activity of Early Embryological Stages 7-31
- Jaffe, 1966; Robinson and Jaffe, 1975; Weisenseel *et al.,* 1975; Novak and Bentrup, 1973

7.19 Multigenerational Study in Rats 7-32
- Coate *et al.,* 1970

7.20 Behavior and Physiology of Dogs 7-33
- Coate *et al.,* 1970

7.21 Mutagenesis in *Drosophila* 7-33
- Coate *et al.,* 1970
- Mittler, 1972

7.22 Mutagenesis in *E. Coli* 7-34
- Coate *et al.*, 1970

7.23 Perception in Several Species. 7-35
- Coate *et al.*, 1970
- Milburn, 1971

7.24 Regeneration. 7-36
- Sisken and Smith, 1975
- Smith, 1974
- Becker, 1972c; Becker and Spadaro, 1972

7.25 Electric Field Effects on Amoeba 7-38
- Friend, 1972; Friend *et al.*, 1975

7.26 References 7-40

Chapter 8—SPECIAL TOPICS

8.1 Cells in Contact: Electrical Properties 8-1
- Loewenstein and Kanno, 1964; Wiener *et al.*, 1964; Loewenstein, 1966; Dewey and Barr, 1962; Azarnia *et al.*, 1974; DiCaprio *et al.*, 1975; Garfield and Daniel, 1974; Spanswick, 1972

8.2 Electrical Cataracts. 8-2

8.3 Air Ionization. 8-2

8.4 Surveys of Farmers and Exposed Persons 8-5
- Busby *et al.*, 1974
- Ware, 1975
- Strumza, 1970

8.5 Electric Field Effects on Plant Life 8-6

8.6 Magnetic Fields in the Human Body 8-7
- Cohen, 1973; 1975a; 1975b

8.7 Orientation of DNA in a Magnetic Field 8-9
- Maret *et al.*, 1975

8.8 Biological Effects of Large Magnetic Fields. 8-10
- Becker, 1963
- Barnothy, 1964; 1969
- Vyalov *et al.*, 1964; Vyalov, 1971
- Mykhaylovskiy *et al.*, 1969; Yashina, 1972; Shishlo *et al.*, 1966; Degen, 1971
- Edlinskiy *et al.*, 1969; Kugot *et al.*, 1969; Odintsov, 1965; Tarakhovskiy *et al.*, 1971
- Vasil'yev *et al.*, 1971a
- Sudakov and Antimoniy, 1973; Dorfman, 1971; Shishlo, 1971

- Pavlovich, 1971; Kogan *et al.,* 1971; Novitskiy *et al.,* 1971; Toroptsev *et al.,* 1971
- Kholodov, 1971
- Ukolova and Kvakina, 1971
- Kholodov, 1970; 1971
- Presman, 1974

8.9 Proposed Mechanisms for Biological Effects by Magnetic Fields 8-20
8.10 Regulations on Electric Field Exposures in the Soviet Union. 8-21
8.11 Regulation of Magnetic Field Exposures 8-24
8.12 References 8-25

Chapter 9—TABULAR SUMMARY OF PRINCIPAL INVESTIGATIONS

9.1 Table 9-1: Electric Field Effects Measured by Total Current Flow 9-1
9.2 Table 9-2: Electric Field Effects Measured by Current Density 9-3
9.3 Table 9-3: Electric Field Effects Measured by Electric Field Strength 9-5
9.4 Table 9-4: Magnetic Field Effects (below 200 G). 9-12
9.5 Table 9-5: Effects in Strong Magnetic Fields (above 200 G) 9-16
9.6 References 9-19

Chapter 10—GENERAL BIBLIOGRAPHY

FOREWORD

This review was undertaken in response to an apparent need for a comprehensive review of the world's literature on the biological effects of low frequency electric and magnetic fields. To undertake this task, our Institute was fortunate in being able to attract Dr. Asher Sheppard to work on this project as a National Institute of Environmental Health Sciences postdoctoral fellow, beginning in October, 1974.

We did not know when this project began that the subject would become a controversial one, and that by the time of the report's publication a number of public interest groups and state agencies would be intervening to block construction of high voltage electric transmission systems because of their concerns about the possible public health effects of the associated electric fields. In addition, there has been concern about the health and ecological implications of the Navy's low frequency communication system, Project Seafarer (formerly Project Sanguine). We hope the information in this report will provide an informational base that will help to approach these issues in an orderly manner.

It would have been satisfying to be able to state that this review, which covers several hundred publications from many countries, and from the journals of many disciplines, enables one to describe exactly and unequivocally the health implications of human exposure to low frequency electric and magnetic fields. Regrettably, this is not possible: there is no evidence that the public health or ecological systems have been jeopardized in the slightest by artificial electromagnetic fields, but there are still stones unturned, and additional studies must be undertaken before it will be possible to state with finality that the matter is closed.

This review was intended to be as comprehensive as necessary for an understanding of the present state of knowledge. To assure both completeness and objectivity, research reports are included that are overly speculative or of less than satisfactory quality. Some readers may wish that we had omitted such references but, in the interest of completeness and objectivity, no such exclusions were permitted.

Merril Eisenbud
New York University
Medical Center

January 15, 1977

CHAPTER 1

INTRODUCTION

Until about one century ago, human exposure to external electric and magnetic fields was limited to the natural fields arising from atmospheric electricity and geomagnetism. However, within all organisms are endogenous electric fields and currents that play a role in the complex mechanisms of physiological control such as neural and neuromuscular activity, tissue growth and repair, glandular secretion and cell membrane function. Because of the role of electric fields and current flow in so many physiological processes, it is not surprising that questions have arisen concerning possible effects on biological systems exposed to artificially produced fields such as those in the transmission and use of electric energy.

Apart from the question of possible human effects is the question of behavioral or other influences on lower animals and plants. There is convincing evidence of field responses in some animal species: certain birds are responsive to the orientation of the earth's magnetic field and, in the case of some fish, electric fields are used for the detection of prey, navigation, and communication.

This report is a review of the existing scientific literature on the effects of electric and magnetic fields from 0 to about 300 Hz, which spans the frequencies at which electrical power is generated, transmitted, or used.

The literature reviewed originates from two sources. First, because electric and magnetic fields are of such fundamental physiological importance, many investigators have studied electrical activity in various organisms to understand a particular life process. These studies have produced an extensive body of literature scattered through the journals of many disciplines (brain research, physiology, neurobiology, muscle research, embryology, bone research, biophysics, behavior, and fish biology).

Second, the subject of the effects of low frequency fields has recently attracted attention because of the possibility of deleterious effects on humans or lower life forms. The major source of environmental exposure is the electrical industry and in addition, interest and research are being generated by the Navy's proposed Project Sanguine (more recently known as Seafarer), a system designed to permit communications with submerged nuclear submarines. The long wavelength at which the Sanguine transmitter will operate (4,000 - 7,000 km) requires a large, buried antenna leading to concern about ecological effects on the considerable land surface influenced by the antenna currents and fields. In 1968 the Navy sponsored a review of extremely low frequency (ELF) research which disclosed such a paucity of relevant research that the Navy undertook its own program of studies. Beginning in 1970, the research has been performed in government and university laboratories, and on the site of a Sanguine test facility in Wisconsin.

The first studies of the effects of live line maintenance of 345 kV transmission lines were begun by the American Electric Power Service Corporation in 1965. A number of studies were undertaken which quantified the levels of exposure, developed calculational and measurement techniques, examined the effects of exposure on laboratory mice, and reported on the results of 9 years of exposure on the health of a small group of linemen. In the U.S.S.R., studies were conducted on the health of workers at a 500 kV substation in 1966, and were first reported in the West at a conference in 1972.

The construction of 765 kV transmission lines is stimulating public interest to the extent that uncertainty about the possible risk from exposure to electric and magnetic fields is presented as a basis for opposition to power line construction in the State of New York.*

In this review, emphasis is placed on 50 and 60 Hz electric fields of: 1) several thousand volts per meter (V/m) and magnetic fields less than 1 gauss (G), the range of exposure near electric power transmission lines; and 2) electric fields less than a few hundred volts per meter and magnetic fields greater than 1 G, typical of exposures from electric appliances and machinery. The future development of the DC, cryogenic, or superconducting transmission technologies may bring about more interest in DC fields and high magnetic fields.

This introduction provides a convenient opportunity to discuss the possible effects of the low frequency fields from a theoretical

---

* State of New York Public Service Commission (PSC) Case Nos. 26529, 26559 concern applications by the Power Authority of the State of New York (PASNY), Rochester Gas & Electric Corp., and Niagara Mohawk Power Corp. to construct and operate 765 kV transmission lines. Cases 26758, 26462 concern an application by Long Island Lighting Co. (LILCO) to construct and operate 345 kV lines. In the 765 kV case, the PSC is conducting generic (rule-making) hearings, beginning in winter, 1975. The intervention by the New York State Department of Health, Department of Agriculture and Markets, the Department of Environmental Conservation, and the Attorney General, is on the basis of the issues of "noise and ozone produced as well as the electrostatic, electromagnetic, and other induced effects of 765 kV lines and the relationship of all of the foregoing on living organisms, including any possible effects on human beings." In the LILCO case, the towns of Huntington and Islip (Suffolk Co.) intervened against the proposed line which would be constructed on an existing, narrow, right-of-way passing through populated areas

point of view. The whole subject would be greatly simplified if the kinds of effects to be expected could be identified from theoretical considerations. However, it will be seen that this is difficult in the present state of knowledge. The absence of a rationale for the assumption of effects is a serious impediment to the design of productive biological experiments.

A low frequency electric or magnetic field may affect living systems by the direct interaction of the field with a biological subsystem and/or by the induction of currents. Biological structures at all levels of organization have electric multipoles and therefore, in principle, can be affected by an electric field. More significantly, their orientation with respect to one another may be altered, and biological molecules can be polarized. These properties, in general, are important functional characteristics of the system. Magnetic multipole moments are so small that -- with the possible exception of some iron-bearing molecules placed in strong fields -- the interaction energy is much less than for the electric interaction.

Both electric and magnetic interaction energies are much too small to affect covalent molecular structure or electrostatic bonds between molecules, but the small amounts of orientational energy may cause steric changes in molecules located at the cell surface.

AC electric and magnetic fields both cause the flow of currents in the body. The electric field within the body, though much weaker than the original field, represents the flow of current in the conducting body tissues. Magnetic fields pass through the body as if it were transparent, but induce currents which flow in circuitous paths. These currents, like those due to an electric field, can possibly interfere with the transmission of information in the nervous system or other tissues, though not by competition with the spikes of the action potential in nerves.

A cellular response to a field implies communication across the cell membrane. This complex region, whose structure and function are imperfectly understood, selectively transmits biochemicals to the cell interior. A strong electric field of about 10^7 V/m is found at the cell membrane where a static potential difference of about 50 mV exists across a distance of about 75 angstroms. Thus, the cell membrane is highly polarized, and the molecular details of its polarization are related to its function, suggesting to some investigators that slight structural changes due to weak applied fields may affect membrane function and allow communication with the cell. There is a need for a more basic understanding of the effects of electric and magnetic fields at the molecular and cellular levels. More complete comprehension of these effects would provide important insight into the mechanisms involved at higher levels of organization.

The most persuasive evidence that organisms are in fact sensitive to electric and magnetic fields involves human and animal behavioral responses in the laboratory, and remarkable observations of fish and birds in nature. A considerable portion of the existing literature, and hence a major part of this review, is concerned with behavioral effects, but an understanding of the underlying biophysical mechanisms is usually not at hand.

The literature reviewed in this report involves diverse biological systems, uses various methods of electric or magnetic field exposure, and studies many different responses. Although there are cases where independent investigators have come to similar conclusions (e.g., implying the importance of the extended cell membrane, or discovering effects on the central nervous system), the overall picture remains incoherent. In part, the problem results from important differences in experimental design which prohibit direct comparisons of the findings of different investigators. More work of high quality is required to augment the evidence that has resulted from several outstanding experimental inquiries. Unfortunately, some research is of poor quality, many reported effects are of dubious statistical significance, and there is little replication.

1.1 NOTE ON THE UNITS USED IN THIS REPORT

In this report, electric field strength is given in units of volts per meter (V/m). The proper MKS unit for magnetic flux density (B) is the tesla (T, formerly the "weber per meter-squared") but, in keeping with the style of most investigators, we use the gauss (G), a CGS unit, which is equal to 10^{-4} T and more convenient (because of the size of the magnetic fields). The CGS unit of magnetic field intensity (H), the oersted (Oe), is often found in the literature. The gauss and oersted are equivalent when the magnetic permeability is unity, as for biological materials; in this case, B and H are numerically equal in the CGS system. Electric current is measured in amperes (A) and potential difference in volts (V). The unit of capacitance is the farad (F), and of frequency, the hertz (Hz, formerly the "cycle per second"). Distances are given in meters (m), centimeters (cm) or angstroms (Å, equal to 10^{-10} m).

1.2 NOTE ON BOOKS, REVIEW ARTICLES AND BIBLIOGRAPHIES AVAILABLE IN THE LITERATURE

A number of prior review articles and bibliographies have appeared in increasing numbers in the last few years. A recent volume, "Biologic and Clinical Effects of Low Frequency Magnetic and Electrical Fields," edited by J. Llaurado, A. Sances, and J. Battocletti, Charles C. Thomas, Springfield, Ill. (1974), includes review articles and research reports. A second recent book, somewhat more tangentially related to the subject of this review, but nonetheless valuable, is "ELF and VLF Electromagnetic Field Effects," edited by M. Persinger, Plenum Press, New York (1974). This volume is more directly concerned with effects due to changes in the natural electrical and magnetic field environment, but the overlap with the fields we are concerned with here is considerable. "Electromagnetic Fields and Life," edited by A. Presman, translated by F. A. Brown, Plenum Press, New York (1970), reviews the effects of electricity and magnetism over a large portion of the frequency spectrum, with most emphasis on effects at microwave frequencies.

Papers from a New York Academy of Sciences conference held in 1973 on "Electrically Mediated Growth Mechanisms in Living Systems," Ann. N.Y. Acad. Sci. 238: 530-542 (1974) were issued in a volume edited by A. Liboff and R. Rinaldi. The papers are organized into three categories: physical and chemical studies, biological studies, and clinical studies. Most of the research is with static fields, but is presented in a manner that makes this collection valuable in understanding the influence of electrical currents or fields on biological systems.

An encyclopedic bibliography was prepared for the Electric Power Research Institute (EPRI) by J. Bridges of the IIT Research Institute, "Bibliography on Biological Effects of Electric Fields," EPRI Research Project 381-1, Electric Power Research Institute, 3412 Hillview Ave., Palo Alto, Calif. (1975). This review includes an index which organized the articles according to the type of work prominent in the research report. For example, effects are categorized as behavioral, biochemical, biophysical, or engineering, and its frequencies are divided up according to DC, subaudio, audio, radio, or microwave regions of the spectrum.

A recent bibliography (untitled) has also been compiled by the Bonneville Power Administration, Box 3621, Portland, Ore. (1974); and the Environmental Protection Agency has compiled a bibliography, "A Partial Bibliography Relating to Health and Environmental Effects of Extremely Low Frequency Electromagnetic Radiation," U.S. Environmental Protection Agency, Environmental Analysis Division, Washington, D.C. (1975).

A "Bibliography of the Biological Effects of Static Magnetic Fields," compiled by L. Gross, is included in the Book "Biological Effects of Magnetic Fields," edited by M. Barnothy, Plenum Press, New York (1964). An older bibliography entitled "Bibliography on the Biological Effects of Magnetic Fields," edited by A. Jacobs, appeared in the Proceedings of the Federation of Experimental Biology Societies 21 Supp. 12: 1 (1962).

The Medical Research and Development Command in the Navy has compiled a bibliography of "Reported Biological Phenomena (Effects) and Clinical Manifestations Attributed to Microwave and Radio Frequency Radiation," edited by Z. Glaser, and available from National Technical Information Service (NTIS), Springfield, Va. (1970), as AD 750 - 271 and Supplement 4 , AD 770 - 621. A fifth supplement was published in 1974 (AD 784 - 007), a sixth in 1975 (AD A015 - 622), a seventh in 1976, and the eighth supplement is in preparation.

The Navy-sponsored research for Project Sanguine is collected in a "Compilation of Navy-Sponsored ELF Biomedical and Ecological Research Reports," EMRPO-2, Electromagnetic Radiation Project Office, Naval Medical Research and Development Command, U. S. Navy, Bethesda, Md. (Feb., 1975), Revision 1 (May, 1975), and is available from NTIS, Springfield, Va. as AD-015 068/OWP and AD-015 069 8WP. A wide range of items selected from the literature are included in Annex C to the "Sanguine System Final Environmental Impact Statement for Research, Development Test and Evaluation," Naval Electronic Systems Command, U. S. Navy, Bethesda, Md. (April, 1972), and available from NTIS, Springfield, Va. as PB 199-732 F2 (1972).

The Institute Recherche de l'Hydro-Québec (IREQ), Québec, Canada has issued a report, "Certain Ecological Effects of High Voltage Power Lines," prepared by N. Hyltén-Cavallius, IREQ-1160 (1975). The text discusses the interaction of electric and magnetic fields with the body in engineering terms and also reviews a selected number of the research reports listed in the bibliography. In addition to other topics, the results of an experiment on the human perception of capacitive discharges are presented.

A recent research report for Project Sanguine (Gavalas-Medici and Magdaleno, 1975) begins with a review of the behavioral studies of electric fields in the low frequency, low voltage regime ("An Evaluation of Possible Effects of 45 Hz, 60 Hz and 75 Hz Electric Fields on Neurophysiology and Behavior of Monkeys," Univ. of California, Los Angeles, ONR Technical Paper").

Previously, the psycho-physiological research on low frequency fields was reviewed extensively (Persinger *et al.*, *Percept. Mot. Skills* 36: 1131, 1973) with a number of references to biological, physiological, and biochemical changes as well.

A collection of Russian and Western references in the magnetic field literature is given in "Influence of Magnetic Fields on Biological Objects," edited by Y. Kholodov, Moscow, 1971, and available in English translation from NTIS, Springfield, Va., as JPRS 63038 (1974). Additional references to Russian literature (mostly at higher frequencies) are given in "Central Mechanism of Action of Electromagnetic Fields" by K. Sudakov and D. Antimoniy, also available in English translation from NTIS as JPRS 60711 (1973).

L. Young's book, "Power over People," Oxford University Press, New York (1973), describes the events surrounding construction of a 765 kV transmission line in rural Ohio and its esthetic and cultural impact, and presents several reasons for opposing such lines. The line's safety is questioned because of possible hazards from ozone production and electric fields, and the latter concern is amplified in the epilogue to the paperback edition of the book (Oxford University Press, New York, 1974). This book gives insight into the reason for contemporary popular concern about the construction of transmission lines and is written for the perspective of the layman.

1.3 NOTE ON REFERENCES FOLLOWING CHAPTERS

Items marked by a dagger (†) are included in the "Compilation of Navy-Sponsored ELF Biomedical and Ecological Research Reports," EMRPO-2, February, 1975. Revision 1, May, 1975. Revision 2, August, 1975. Electromagnetic Radiation Project Office, Naval Medical Center, Bethesda, Md. 20014. The original report (February, 1975) is available from NTIS as AD A015 068/OWP and AD A015 069/8WP.

Reference items indicated as "available from NTIS" may be ordered from: Dept. of Commerce, National Technical Information Service (NTIS), 5285 Port Royal Road, Springfield, Va. 22161.

Special Publication #10 of the IEEE Power Engineering Society, a translation by G. G. Knickerbocker entitled "Study in the U.S.S.R. of Medical Effects of Electric Fields on Electric Power Systems," may be obtained by writing: Single Publication Sales Department, IEEE, 445 Hoes Lane, Piscataway, N.J. 08854.

CHAPTER 2

SUMMARY

2.1 FIELDS IN THE ENVIRONMENT

An understanding of the electric and magnetic fields of natural origin provides a logical baseline for a discussion of the man-made low frequency* fields to which people are exposed. In fair weather the DC atmospheric field is 120 - 150 V/m and may reach 10,000 V/m during an electrical storm. The DC geomagnetic field is about 0.5 G and relatively stable. Both fields drop rapidly in intensity over the ELF region to values of 10^{-4} V/m and 10^{-8} G at a frequency of 50 - 60 Hz (Polk, 1974).

Man-made 60 Hz fields, measured 30 cm from a variety of home appliances, were found to vary from 10 - 250 V/m and from 0.001 - 25 G (D. A. Miller, 1974). Ambient levels in typical homes were measured in the range of 1 - 10 V/m and 0.001 - 0.1 G. For example, the electric field 30 cm from a light bulb is only 2 V/m. Fields of 60 V/m were measured near a refrigerator, and 250 V/m at 30 cm from an electric blanket. The magnetic field associated with a refrigerator is very low, less than 0.01 G, but the magnetic field near a color television receiver was measured at greater than 1 G and near a hair dryer the magnetic field was more than 10 G. (See Tables 3-1 and 3-2.)

2.2 FIELDS IN THE BODY

The presence of a body in an otherwise uniform electric field distorts the field so that the field strength at the surface of the body is markedly increased. The surface field intensities depend on the radius of curvature and orientation in the field so that, for a human body standing in a vertical field, maximum intensities occur at the head. Alternating currents flow through the body, which is a relatively good conductor, in a direction parallel to the electric field, but the exact current paths are unknown.

Near ground the total current flow through the body, I is determined by the field strength, E; body capacitance, C; and effective height above ground, h. The product E • h gives the potential, V_o, to which the body would be raised if it were perfectly insulated from ground. The body current is,

---

* The frequencies discussed in this report are in the extremely low frequency (ELF) portion of the spectrum, usually taken to include frequencies below 300 Hz. Most of the research reviewed is below 75 Hz and some involves static fields and direct currents.

$$I = \omega V_o C,$$

where $$V_o = E \cdot h,$$

and, ω is the angular frequency of the field. An equivalent form may be written by combining the effective height and capacitance into an equivalent area, A, so that,

$$I = \omega \varepsilon_o E A,$$

where ε_o is the dielectric permittivity of free space (see Figure 2-1). The capacitance to ground (and thus the effective area) is maximized when the person is standing on the ground and minimized when he is suspended in space far from the ground.

In this latter case, the above formulae cannot be used to calculate the currents because the capacitance to ground is not the dominant capacitance and thus the potential cannot be described just by V_o (See Figure 2-2). Thus, for a body held in space far from the ground, the current through the body is determined by the potential distribution about the body's surface and by the body's capacitance to free space. The calculation involves only a solution of Laplace's equation to get the proper potential distribution on the surface. These surface potentials determine the electric field at the surface and the resulting internal current flow. This problem has been solved in spheroidal coordinates and the results are available in tabular form (Barnes *et al*., 1967). The capacitance to free space depends on the body's size and shape. Using a sphere of radius R to approximate the body, its capacitance far from ground is given by,

$$C = 4\pi \varepsilon_o R,$$

giving a capacitance of about 50 pF for R = 0.5 meter.

A third situation arises when the person is in contact with the conductor as in live-line maintenance work. In this case, the body currents can be calculated using a numerical technique described by Barnes *et al*. (1967).

In the first case (Figure 2-1) we can discuss, for example, a man standing on the ground in a 10 kV/m, 60 Hz external field. He would be subjected to a current of about 200 μA, an approximate current density of 1.6 mA/m^2 through the trunk of the body,* and an internal electric field of 8.0 mV/m, assuming an internal resistivity of 5 ohm-meter.

---

* We assume the current is uniformly distributed through the midsection of the stout torso used in the spheroidal model of Barnes *et al*. (1967). Smaller, more realistic cross-sections increase the mid-torso current density two or three times.

Figure 2-1. Current flow through the human body located near ground in an electric field

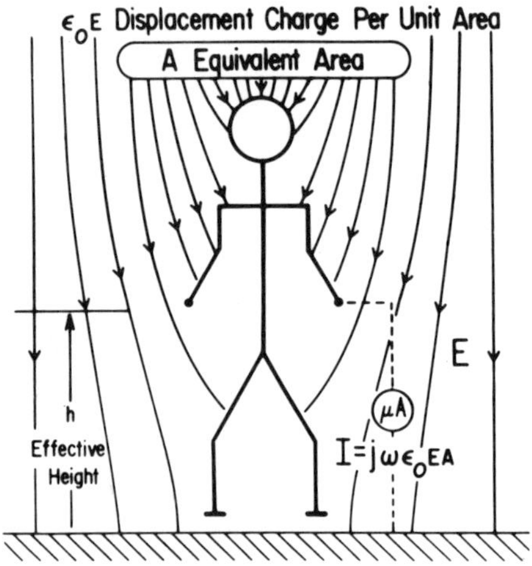

Figure 2-1a.

Schematic representation of a man, near ground, in a uniform field, showing field distortion, the man's effective height and area. If the indicated circuit to ground is completed the current I, short-circuited to ground, may be measured by a microammeter.

Source: Deno and Zaffanella, 1975

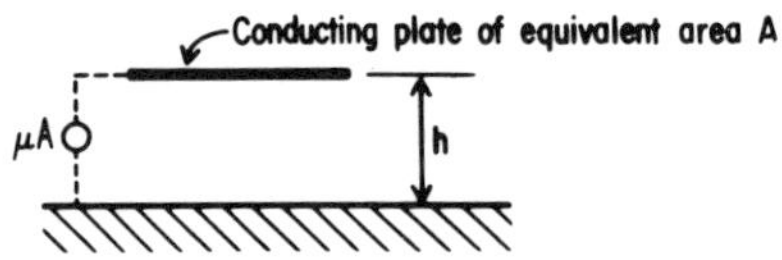

Figure 2-1b.

Electrical representation of a man, near ground, in an electric field.

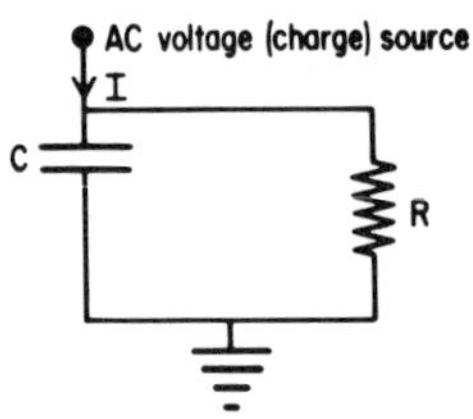

Figure 2-1c.

Equivalent circuit elements: C, total body capacitance; R, resistance to ground.

Source: Deno and Zaffanella, 1975

Figure 2-2. Current flow through the human body located in an electric field remote from ground

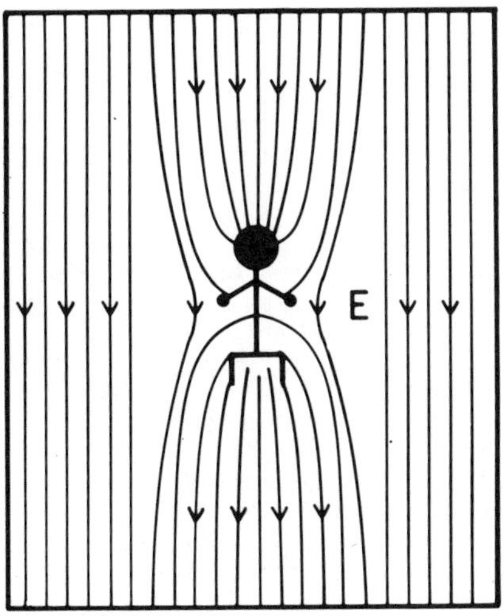

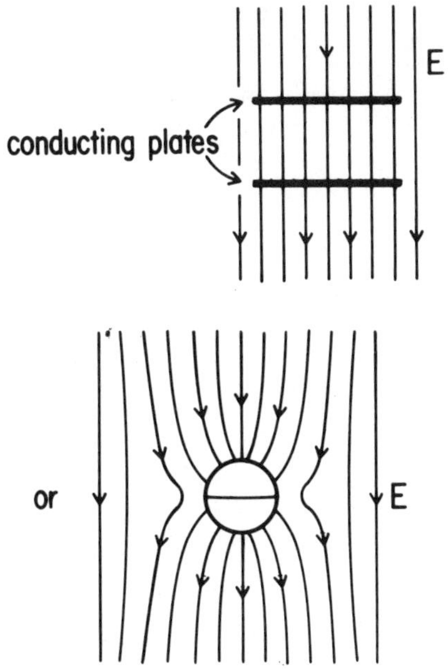

Figure 2-2a. Schematic representation of electric field lines of force.

Figure 2-2b. The analogous representation of a parallel plate capacitor or conducting sphere place in an electric field.

In the second case (Figure 2-2), a man suspended in a uniform field of 10 kV/m, 60 Hz, the total current flow is about 65 μA, the current density through the trunk about 0.52 mAm^2, and the internal electric field 2.6 mV/m, again assuming a resistivity of 5 ohm-meter.

In the third case, a person in contact with a conductor at a potential of 200 kV, the total current is about 3 mA (Barnes et al., 1967). In practice, the lineman is in a bucket which provides shielding and significantly reduces the total current to about 250 μA.

For the purpose of comparison, it may be noted that the currents induced in the body by an external electric field can exceed the minimum perceptible current that can be perceived by touching a finger to an energized conductor (0.1 mA), or by gripping a conductor with the palm (0.5 mA). However, the currents induced by electric fields are not perceived, probably because of the extended area through which the current flows and the consequently small current densities at the skin.

Current is conducted through the body mainly by the ionically rich extra-cellular fluids with little of the current conducted through the cell membrane. Membrane conductivities vary over 3 orders of magnitude, and cell sizes are also quite variable, making it difficult to state a typical value for the transmembrane current. The uncertainty is illustrated by calculations based on different assumptions of reasonable parameters for the cell, which show that from 0.2 - 20% of the current near a cell may enter through the membrane (Schwan, 1972).

Nevertheless, the serious consequences of an electric shock are due to the passage of significant amounts of current into cells, especially in the nerve and muscle tissues of the heart, chest muscles or diaphragm. From studies of electric shock it is known that currents of at least 5 mA are required to produce a shock which is disabling because the victim can't let go of the conductor and a current of 60 mA or more is needed to cause fibrillation of the heart muscle in adult human beings (Keesey and Letcher, 1970). Currents of this magnitude would flow through the body only in fields greater than 250 kV/m (corresponding to the let-go current) and 3 MV/m (corresponding to the fibrillation current). (Fibrillation has occurred when a current of 200 μA is applied directly to the heart using electrodes, but this is not a realistic exposure situation.) By estimation of the fraction of the total body current which affects the heart, Schwan (1972) calculates that the minimum current density in tissue which can cause fibrillation is between 1 and 10 A/m^2. This current density requires external fields of between 6 and 60 MV/m -- fields so large that they exceed the dielectric breakdown strength of air. The possibility of a disabling primary shock among persons in the vicinity of power lines can be dismissed. However, the question of "micro-shocks" is another matter, to which we now turn.

A strong electric field, by definition, requires large potential differences in the space occupied by the field. Thus, an object a meter or so above ground, and electrically insulated from ground, can be at a high voltage with respect to the ground. Such large potential differences make possible electric discharges which are sometimes referred to as "microshocks," "minishocks," "secondary shocks," "spark discharges," or the "discharges due to electrostatic induction." The electric fields near transmission lines can be sufficiently high to cause discharges between the body and objects at a different voltage. The severity of the discharge and accompanying shock sensation depends on the potential difference between the person and the object touched and on the capacitance of the charged object or person. The discharge may be to ground, as in the case of blades of grass which can

Figure 2-3. Secondary shocks (discharges) in strong electric fields

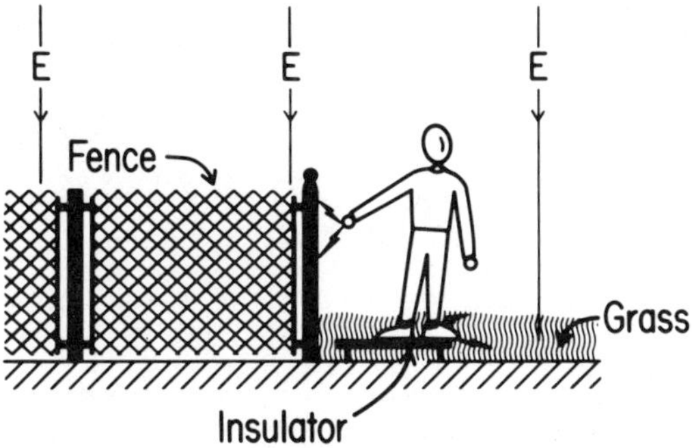

Figure 2-3a.

A man, insulated from ground, is at a high voltage and receives discharges from a grounded fence post or from the grass.

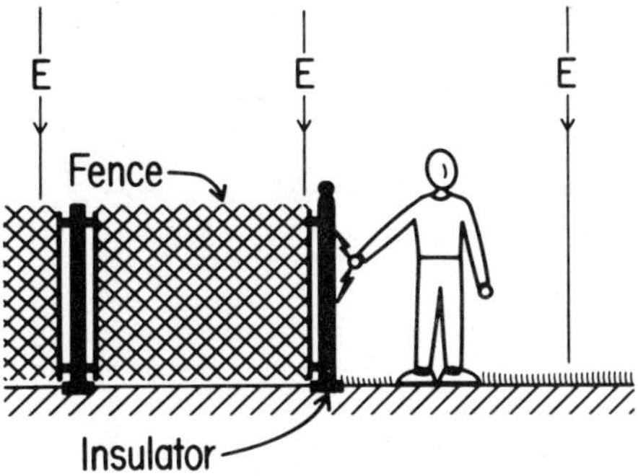

Figure 2-3b.

A man, grounded, receives discharges from an insulated fence.

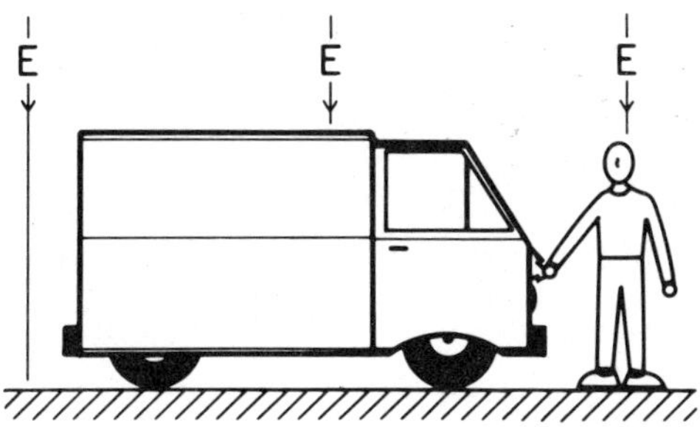

Figure 2-3c.

A man receiving discharge shocks from a car or truck.

cause uncomfortable spark discharges in the vicinity of the ankles, or from a source at higher voltage, as when touching a vehicle which is geometrically higher and insulated by its rubber tires. (See Figure 2-3).

These spark discharges are too small to be life-threatening, but can be annoying and possibly a cause of accidents. There is also evidence from changes in pulse, respiration and the electroencephalogram (Krivova, 1968) that these discharges could have significant physiological effects (apart from the direct action of the electric field).

Unlike electric fields, magnetic fields penetrate the human body, which has about the same magnetic permeability as free space. AC magnetic fields can induce currents that will flow in closed circuits within the body. It can be calculated that a 1 G, 60 Hz magnetic field induces a 2 mV/m electric field in a circuit of radius 0.1 meter, composed of material having the same resistivity as body tissues. This internal field produces a current density that is comparable to the one produced by a 2.5 kV/m external electric field. Thus, for sufficiently large magnetic fields, the induced current flow can be comparable with that produced by electric fields. However, the current pathways are not the same and the magnitude of the magnetic field-induced current is proportional to the area enclosed by the circuit. In any case, the ground level magnetic fields in the vicinity of power lines are not strong enough to cause currents comparable with the electric field-induced currents, although the reverse may be true in the vicinity of certain home appliances and industrail equipment where magnetic fields can exceed 10 G.

The direct interaction of a magnetic field with the cells and molecules of biological tissues involves energies that are very much smaller than thermal energies. For this reason, a direct physiological effect due to heating or the disruption of chemical bonds in impossible. Most biological molecules are diamagnetic and some exhibit paramagnetism due to the presence of free radicals, iron atoms, or atoms of the transition elements. Physiological effects involving these magnetic properties are possible for fields greater than several hundred gauss and may be important at smaller field strengths.

2.3 FIELDS INVOLVED IN NORMAL BIOLOGICAL FUNCTIONS

The importance of DC or ELF electrical activity in normal biological functions is illustrated by the role of electric or magnetic fields, or their associated currents, in the following phenomena.

1) The normal brain rhythms are observed by electroencephalogram and magnetoencephalogram patterns which are predominantly below 20 Hz, although higher frequencies are present. The cooperative manner by which these rhythms are established and the nature of their interaction with individual brain cells is not yet determined.

2) The activity of the nervous system, in which neurons propagate direct current pulses along paths of varying length, creates electrical signals which affect muscle or gland function. Such currents may also represent sensory information, or even the "stuff" of thought itself.

3) Bone growth, and the regeneration of new bone after injury, are associated with DC (and perhaps AC) electric currents. Because the protein constituent of bone, collagen, is piezoelectric, electrical potentials are created when a bone is subjected to stress. These potentials stimulate growth, but the manner in which the electrical potentials regulate normal bone growth at the cellular level is not yet known.

4) There is a very intense (about 10^7 V/m) electric field located at the cell membrane due to the potential difference maintained between the inside and outside of the cell. The membrane barrier is surrounded by a more diffuse region where biomolecules are active in identifying other molecules in the cellular environment. Scientists are currently attempting to discover the mechanisms and functions localized to this region, where recognition of hormones and enzymes occurs, according to the structural details of the molecules. The structural details can, in principle, be expressed as electrical or electronic properties, but a quantitative description of the effects of imposed fields on molecular structure is not at hand.

5) Natural sensitivities to very weak electric and magnetic fields have been demonstrated in fish, birds, and even bacteria. Migrating birds or fish use the naturally-occurring external fields for navigation and some fish use electric fields to sense objects, or for communication. The electroreceptor in some fish species has been identified, but the magnetoreceptor in birds is not known. A recent report describes the response of a mud-dwelling bacterium to the direction of the earth's magnetic field (0.5 G).

6) DC potentials and a high degree of intercellular (electrical) contact characterize the cell mass during early states of embryonic growth, and are associated with the process of differentiation at its very beginning. This research, conducted in plant and animal embryos, implies an important role for electrical signals in regulating cell processes which are ultimately directed by the cell nucleus.

7) DC potentials between different parts of the body may be measured on the skin surface, although their interpretation has not been established. DC potentials exist between portions of the skull and along the spinal column as well as along the surface of the skin of the limbs and torso.

2.4 FIELD EFFECTS IN ANIMALS AND HUMANS

Biological effects due to the imposition of external fields have been seen in a large number of systems over a wide range of field strengths and ELF frequencies. The studies performed on animals and human subjects can be categorized into 4 areas: a) neurophysiological and behavioral; b) biochemical; c) clinical, physiological and behavioral; and d) reproductive.

Neurophysiological and Behavioral Effects

Both electric and magnetic fields of relatively low strength appear to affect higher functions of the brain involving cognition or perception of the passage of time. A series of experiments, performed over a 6-year period at the Brain Research Institutue at the University of California at Los Angeles, exposed monkeys to ELF electric fields at 7, 10, 45, 60, and 75 Hz, using field strengths between 1 and 100 V/m. The first experiment (Gavalas *et al.*, 1970) at 7 V/m, 7 Hz, showed a field-related 0.4 second change in the inter-response time (IRT) of monkeys previously conditioned to respond at intervals of about 5 seconds. No effect was demonstrated for a 7 V/m electric field at 10 Hz. Later, a more complete experiment (Gavalas-Medici and Magdaleno, 1975) again demonstrated changes in the monkey IRTs, but the effect at 7 Hz, 10 V/m is much smaller (about 0.12 second). Experiments conducted at other frequencies and field strengths showed greater sensitivity at lower frequencies and suggested that there may be a long period before the effects wear off; this was especially noticeable in the external field of 100 V/m. (See Tables 7-2 and 7-3 for representative data and Figure 7-1.)

These interesting behavioral results are now supported by *in vitro* studies on chick and cat brains which show altered rates for the efflux of calcium ions and GABA* from brain tissues exposed to various forms of electrical energy. In one experiment (Bawin and Adey, 1975), a 20-minute exposure of chick brain to 10 and 56 V/m electric fields at 6 and 16 Hz reduced the efflux of calcium from the brain tissue; but the effect was absent for fields at 1 or 32 Hz and for all frequencies at 5 V/m. For cat brain, the calcium efflux reduction was seen at 56 V/m (16 and 6 Hz) but no significant effects were seen at lower field strengths or at 1, 32 or 75 Hz. These results suggest a threshold at below 10 V/m (chick brain), or below 56 V/m (cat brain), and a peak response frequency at about 6 Hz. This is in general agreement with the behavioral experiments on monkeys described above (Gavalas *et al.*, 1970; Gavalas-Medici and Magdaleno, 1975) where a threshold at about 10 V/m is suggested and the maximum effects occur at 7 Hz.

In an earlier experiment Kaczmarek and Adey (1974) showed an *increase* (c.f. above) in the efflux of calcium and GABA from cat

---

* GABA (γ-aminobutyric acid) is an inhibitory neurotransmitter.

brain tissue exposed to a pulsed current supplied by electrodes (non-metallic). The 1 millisecond pulses, repeated 200 times per second, caused an internal field in the brain tissue of from 2 - 5 V/m, a much greater internal field than in the other cases.

Other experiments on the efflux of calcium ions from brain tissue have been conducted using an ELF modeulated VHF radio wave (147 MHz) (Bawin _et al._, 1975). These results demonstrate an increased efflux and also indicate frequency discrimination with a peak response at a 16 Hz modulation frequency (See Figure 2-4).

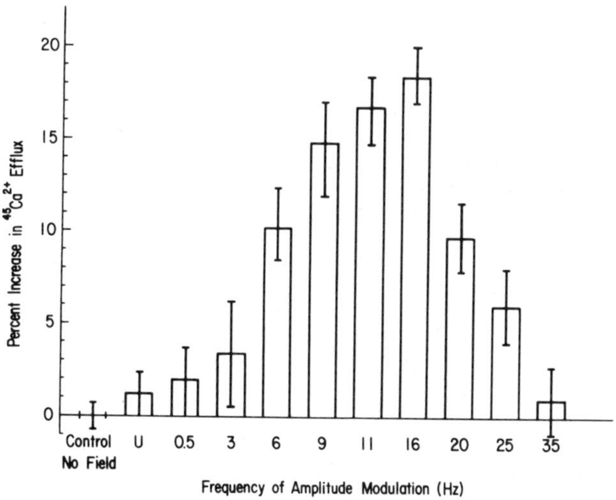

Figure 2-4. Ca^{++} efflux from chick brains exposed to an ELF modeulated VHF signal.

Source: Bawin _et al._, 1975.

Earlier experiments showed that the modulated VHF field had a marked positive effect on the performance of conditioned behavior in cats (Bawin _et al._, 1973). In addition, there was a remarkable 5-fold increase in resistance to the extinction of the trained behavior and EEGs showed changes in the brain rhythm. However, no brain wave changes were seen in the absence of the conditioning procedures, indicating that the external field acted as a "reinforcer."

There was some question about the possibility of artifacts due to an "antenna effect" caused by the EEG leads. This is unlikely because of the careful shielding of the leads, the anatomical localization of the detected signals, the transient nature of the patterns, and their frequency specificity. In one study (Gavalas-Medici and Magdaleno, 1975) the same behavioral effects were observed in the two monkeys without EEG leads as in the 3 bearing leads.

These results are not fully understood but Adey (1974) has proposed a means by which the extended membrane region surrounding the neural membrane may be sensitive to electric charge movement or to electric fields. This region, perhaps 2000 angstroms thick, (c.f., the neural membrane itself which is about 75 angstroms), is referred to as the glycocalyx. Receptors for the hormones and neurochemicals which are effective in very small concentrations are located in this region. The extraordinary sensitivity of the glycocalyx to chemical and electrical signals is explained by the hypothesis of a membrane amplifier which utilizes metabolic enzymes (e.g., adenyl cyclase) to release large amounts of energy by changing ATP to c-AMP in response to a small input signal that may be represented by a conformational change in the molecules at the receptor site.

Because the calcium ion (Ca^{++}) binds more strongly to the macromolecules of the glycocalyx than any other ion, excepting the hydrogen ion (H^{+}), it may play a key role in the mechanism of membrane sensitivity. Ca^{++} binding occurs at the numerous sites of negative charge found on the macromolecules, leading Adey to propose that a cooperative conformational change occurs over large distances along the membrane, thus releasing large amounts of energy as the calcium ions make small, local movements in response to the new configurations.

These ideas are in accord with current knowledge of membrane chemistry and for the time being must remain speculative. The suggestion that the membrane's susceptibility to very small external fields occurs by the mechanism of conformational, cooperative changes in the macromolecules of the glycoalyx is similar to the current explanation of the potency of a very low drug and hormone concentration. In both cases, the conformational changes act in conjunction with a biochemical amplifier which uses the small, trigger-like changes in macromolecules to liberate much larger amounts of energy.

This mechanism is independent of intracellular metabolism, as indicated by experiments using ELF-modulated VHF fields in which calcium efflux was stimulated in chick brain cells poisoned by potassium cyanide (Bawin _et_ _al_., 1975).

In this and other experiments using a VHF wave modulated at extremely low frequency, the results are attributed to demodulation of the VHF carrier at the surface of the glycocalyx where the localized excess of fixed negative charges (anions) is said to give the region rectifying properties (Adey, 1974).

Although these experiments indicate the peak sensitivity is in the range of 6 - 16 Hz, there may still be response at higher frequencies as the experimental results at 60 and 75 Hz indicate. Apparently, greater field strengths are required at the higher frequencies (see Tables 7-2 and 7-3). At this time there is no indication of a sharp frequency cut-off. Indeed, these studies are part of a work on neuronal communication in the brain where weaker rhythms are observed at frequencies well above the dominant, low frequencies.

These experiments are unique among the research discussed in this review because of the thoroughness of the research, the extent of replication, and the consistency of the results. Nonetheless, there are inconsistencies among data from different experiments that probably reflect the difficulties involved in behavioral experiments and indicate the need for additional research to supplement the preliminary experiments.

First, the 7 Hz IRT effect changed from 0.4 second in the pilot study (7 V/m) to 0.12 second in the full study (10 V/m), which the authors attribute to increased initial training and better statistics from more data. Second, a significant effect was not seen at 60 Hz, 56 V/m, although results were significant for 56 V/m fields at both 7 and 75 Hz. The small IRT effect (smaller than the standard deviation about the mean IRT) was statistically significant at the 0.05 level (standard error of the difference is smaller than the difference) for 3 test conditions. Third, these 3 cases contrast with 5 other nonsignificant tests above 10 V/m where frequency dependence of the threshold (for data at 10 V/m, 45 Hz and 75 Hz and perhaps at 56 V/m, 60 Hz) and persistence effects (at 100 V/m) may cause the apparent inconsistency. To the contrary, there is consistency in that, whether significant or not, the results are uniformly in the direction of shorter IRTs for the exposed group and the effect appears to increase with field strength (below 100 V/m) (see Table 7-3).

With regard to the hypothesized mechanism there are still many steps to be taken before the existence of conformational changes is established and the details of the biochemical amplifier are known. In particular, there is a need for quantitative evidence to demonstrate the theoretical plausibility of conformational changes by the low level fields ($\sim \mu V/m$) and currents ($\sim \mu A/m^2$) which seem to affect the brain. Finally, it remains to be seen what consequences follow from these changes. Would the effects on brain function be more pronounced at higher field strengths? Why are the behavioral effects demonstrated so far limited to such functions as time perception -- if indeed that is the correct interpretation of the results? And, if these weak fields can cause such substantial effects on Ca^{++} binding, why aren't there more obvious functional effects in organisms exposed *in vivo*?

In humans, cognitive skills were apparently affected by a 24-hour exposure to a magnetic field of 1 G at 45 Hz (Gibson and Moroney, 1974). All the subjects showed *poorer* performance on a test of speed and accuracy in the addition of numbers. But in a task requiring symbol memorization, tests made in the post-exposure period showed that the exposed subjects performed significantly *better* than the controls. Although these effects are statistically significant, they are slight and the authors caution that it would be premature to conclude that central nervous system (CNS) function is affected on the basis of findings in a pilot experiment, especially in view of the other tests in the experiment which indicated that other CNS functions were unaffected.

In guinea pigs exposed for 0.5 hour to a 50 Hz electric field at 100 kV/m, brain electrical activity was altered (Blanchi *et al.*, 1973b). Electroencephalograms (EEGs) of rats exposed to a 10 kV/m DC electric field indicated increased brain wave activity, with a prompt return to pre-test levels when the field was switched off (Lott and McCain, 1973).

In humans, no effect on the EEG was observed in a study using a 15 kV/m field (50 Hz) for a 3-hour period (Hauf, 1974). In another study, a high voltage source was used to produce electric discharges without an accompanying ambient electric field (Krivova, 1968; Krivova *et al.*, 1975a). In this study of humans, effects on the central nervous system were demonstrated by altered EEGs and changes in pulse and respiration rates. In view of the annoying character of spark discharges, it is difficult to interpret whether this report indicates effects beyond those expected in response to an unpleasant stimulus.

Biochemical Effects

Biochemical effects have been demonstrated both *in vitro* and *in vivo*. The *in vitro* studies expose biological materials to unattenuated electric fields that are thus of greater strength than the fields found within the body in the *in vivo* studies; the magnetic field exposures are equivalent in both cases.

Riesen *et al.* (1971) have demonstrated a reduction of oxygen uptake by mitochondria and reduced norepenephrine uptake by synaptosomes (nerve cell organelles) when the mitochondria were directly exposed to a 60 Hz electric current corresponding to a very strong electric field and the synaptosomes to a magnetic field of 50 - 100 G (60 Hz).

In contrast to the magnetic field results with synaptosomes, identical experiments with mitochondria did not show response in the range to 50 - 100 G (60 Hz).

Bassett and Hermann (1968) used a strong electrostatic field (10^5 V/m) to produce an increase in the DNA and collagen production of cultured fibroblast cells. The stimulation of DNA production indicates an effect on the cell nucleus.

These remarkable results bear further investigation; however, the electric fields used were much stronger than those to which cells inside the body are exposed as a consequence of an externally imposed field. In the mitochondria and synaptosome experiments, the magnetic fields (50 or 100 G) were within the range of values observed environmentally.

Using whole body exposures, Marino (1974a) has found an alteration of blood proteins, steroid levels, and a decrease in body weight for rats exposed continuously, for 30 days, in a strong 60 Hz electric field (15 kV/m nominal value). Compared with the controls, the experimentals showed a γ-globulin protein fraction decrease of about 20%, an albumin increase of 7%, and a corticosterone level decrease of about 40% (Table 5-5a). Further experiments (Marino _et al._, 1976a,c -- Tables 5-5b,c,d) show that these statistically significant results are not found consistently but a strong, consistent effect on the exposed animal's water consumption was noted. This finding may indicate a possible artifact from microshocks or microcurrents which occur when the animal drinks, although Marino _et al._ (1976c) suggest that a detailed data analysis argues against an effect at the water bottle. The authors' conclusion is that "quantifiable biological changes" occurred in the serum levels of proteins, steroids and SGOT, in pituitary and adrenal weights and in body weight. These changes are interpreted (Becker, 1974b, Marino _et al._, 1976a) as evidence of a stress response in the animals. Questions of whether the observed changes are influenced by an electrostatic effect (spark discharges), microcurrents, or are purely a field effect should be addressed in further investigations of these important results.

Marino _et al._ (1976b) also report a pilot study on three generations of mice exposed continuously to strong electric fields. The mice exposed to a vertical field at 15 kV/m experienced very high mortality and had much lower body weights at 35 days of age. Mice exposed at 10 kV/m to a horizontal electric field field showed similar but weaker effects (Table 5-6), indicating possible dependence on field strength and orientation, or perhaps the importance of the water bottle problem mentioned above.

Using DC fields for long-term exposures, Marino and coworkers (Marino _et al._, 1974b -- Table 5-7) have shown similar effects on blood proteins and have reported two additional remarkable effects: the occurrence of glaucoma in rats, and chromosome changes in tumor cells exposed to the field after injection in the host animal (Table 5-8). Becker, in whose laboratory these experiments were performed, believes that glaucoma and blood protein changes again indicate stress, but the mechanisms by which the animal is stressed are not known (Becker, personal communication). The chromosome changes are not explained.

Beischer _et al._ (1973) found an increase in serum levels of the triglycerides* 1 - 2 days after the exposure of men in a 1 G magnetic field at 45 Hz (see Figure 2-5). Many other tests were performed on the blood and urine samples of the subjects with negative findings. Because the effect on lipid metabolism appeared one or two days after exposure, Beischer speculates that the field interacts with the chemical precursor of an enzyme, perhaps triglyceride lipase, so that the effects are not seen until the enzyme stores are depleted.

This increase in the serum triglyceride levels in man was not confirmed in rhesus monkeys at 9 G (15 and 45 Hz) (deLorge, 1974). In contrast, the bi-weekly measurements showed a decrease

---

* The new, preferred name for these lipids is the triacylglycerols.

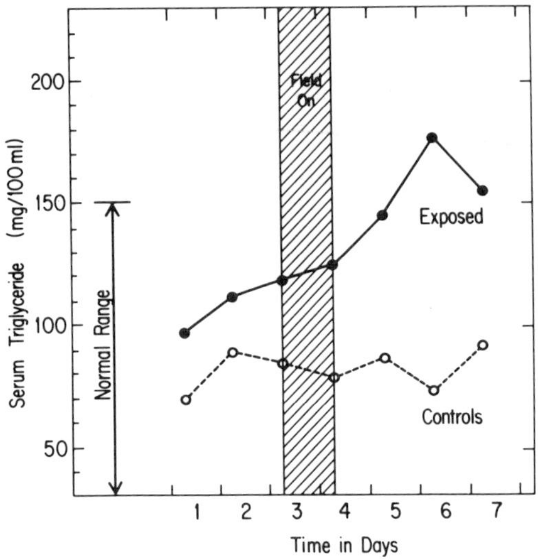

Figure 2-5.

Serum triglyceride levels in men exposed to a 1 G magnetic field at 45 Hz.

The solid line connects the average daily triglyceride values for 4 exposed subjects; the broken line connects the average triglyceride levels for the 4 control subjects. The rise in the triglyceride level which peaks 2 days after the exposure ends _is_ significant, but the difference in absolute levels between the exposed and control subjects is _not_ significant.

Source: Beischer _et al_., 1973

---

in the triglyceride levels which, deLorge suggests, may be related to the use of a partial deprivation diet.

Clinical, Physiological and Behavioral Effects

Asanova and Rakov (1966) reported on clinical studies of 45 employees of a Soviet 400 - 500 kV substation. These examinations were prompted by complaints on the part of the employees after the first several months of substation operation at the new (higher) voltages. The clinical symptoms observed were described as effects on the central nervous system, particularly the autonomic nervous system. Thirty-seven of the workers complained of headache, sluggishness, fatigue and sleepiness; disruptions of the digestive system and cardiovascular system were also mentioned. One-third of the men between 30 and 40 years of age complained that "sexual weakness" developed after 8 - 10 months of work in electric fields.

The medical examinations resulted in a diagnosis of neural pathology in 28 cases. "Functional disruption of the central nervous system" was diagnosed in 26 cases on the basis of instabilities of the pulse (17 cases) and arterial blood pressure (11 cases), excessive perspiration, and fine tremors of the extended hand. Four persons showed more severe vasomotor and secretory disorders and in 4 cases the neurasthenic syndrome was seen. The electrocardiogram showed sinus bradycardia in 14 cases and slowed atrial and ventricular conduction in 10 cases as well as "diffuse changes in the myocardium." The remaining medical findings were digestive tract disease (6 cases), changes in red and white cell populations and moderate retardation of the sedimentation rate. No ophthalmological disease was seen (see Table 2-1).

This report suffers from insufficient data, lack of controls, and the difficulty of interpreting the manifold clinical symptoms. Nonetheless, the report of such widespread problems (37 of 45 had complaints, 28 were diagnosed to have neural pathology) among a young group of workers (39 of the 45 persons were under 40) is striking. Examination of the employees' clinical records for the 2 years before the study was begun disclosed the same neurological findings, implying that the problems were not an artifact of the intensive investigation. Another Soviet investigator reports that the complaints and "undesirable cases" disappeared following the installation of protective screening in compliance with the Soviet rules for electric field exposure (IEEE, 1971; Bourgsdorf, 1972).

The maximum electric fields in a Soviet 500 kV substation are in the range of 15 - 27 kV/m in open areas away from shielding structures (Korobkova et al., 1972). In most areas of the switchyard, the fields are less than 5 kV/m. The Soviet regulations for electric field exposures (IEEE, 1971) are based upon the assumption that both the "direct influence" of the field and the "discharge influence" are significant (Korobkova et al., 1972). The rules require the prevention of all discharges to the body (sparks) whenever the field exceeds 5 kV/m and, in addition, the permissible exposure periods in fields greater than 5 kV/m and less than 25 kV/m are limited. Above 25 kV/m no exposures are permitted and electrical shields are used to reduce the fields below 5 kV/m wherever possible. The extent to which the complaints and clinical findings among the substation workers are due to the fields or discharges has not been determined.

It would be desirable to conduct a similar study of employees exposed to comparable fields in this country.

A second report from the U.S.S.R. (Sazonova, 1967) discussed physiological tests on substation workers which showed that the group receiving greater exposure had a significantly lower blood pressure, greater neuromuscular irritability, increased latent reaction times, and higher error rates in a stimulus-response test. Those who worked inside the buildings as "operators" may be considered as a control group for the "maintenance" personnel, who had greater exposures to electric fields. Viewed in this way, the test performance of the two groups is substantially different; however, a matched control group would have been preferable (see Table 6-1 and Figure 6-1).

Similar, though tentative, results were noted in a pilot experiment in which dogs were exposed to a weak (1 or 2 G) magnetic field, a 10 or 20 V/m electric field, and a 0.5 mA current between electrodes (Coate et al., 1970). The inter-electrode current, it may be argued, represents exposure to a very large

TABLE 2-1. CLINICAL SYMPTOMS REPORTED IN SOVIET SUBSTATION WORKERS*

1. Exposure Data

Total Group —	(41 men, 4 women)	45
Age Distribution —	less than 30 y	10
	30 - 40 y	29
	over 40 y	6
Work Experience — (in substations)	less than 1 y	9
	1 - 3 y	20
	more than 5 y	5

Average Field Exposure —

Maintenance personnel	> 5 h per day
Attending personnel	< 2 h per day

Electric Potentials, fields and currents —

Near energized equip't			Near disconnected equip't		
Potential	Field**	Current	Potential	Field**	Current
maximum					
26 kV	14.5 kV/m	230 μA	4 kV	2.2 kV/m	35 μA
average					
14-18 kV	7.8-10 kV/m	115-125 μA	2 kV	1.1 kV/m	15 μA

(continued)

*Asanova and Rakov (1966)

**See Section 4.5 of this review for the technique of determining these values which are not given in the research report.

TABLE 2-1. CLINICAL SYMPTOMS REPORTED IN SOVIET SUBSTATION WORKERS (cont'd - 2)

2. Complaints Presented —			41
	headache, sluggishness, fatigue, sleepiness		37
	disrupted activity of digestive tract and cardiovascular system		4
	sexual weakness		1/3 of the men between 30 & 40 y
	palpitations, pains in cardiac region, irritability and poor appetite		—
3. Medical Exam Findings —			
	Neural Pathology		28
	Functional disruption of the CNS		
	autonomic dysfunction	(majority)	
	neurasthenic syndrome	6	
	Cardiovascular disease		14
	neurogenic hypertension	4	
	arteriosclerosis	3	
	Bradycardia		12
	Tachycardia		5
	Hypotension		7
	Hypertension		4
	Electrocardiography exam		
	sinus bradycardia		14
	slowed atrial and ventricular conduction		10
	diffuse changes in myocardium		"some"
	Digestive tract disease (chronic gastritis, colicystitis)		6
4. Hematology Findings			
	Moderate Thrombocytopenia Moderate Neutrophilic leukocytosis Leukopenia with neutropenia		"most frequently encountered"
	Lymphocytosis and monocytosis		"also observed"

TABLE 2-1. CLINICAL SYMPTOMS REPORTED IN SOVIET SUBSTATION WORKERS (cont'd - 3)

Changes in red blood cells	"more rarely"
Reticulopenia, lowered hemoglobin and erythrocytes	"tendency toward"
Retarded sedimentation rate	19
Marked spherocytosis	17 of the 28 cases studied

electric field. Hypertension and a possible interference with thermal regulation were found.

Ten American linemen were given medical examinations over a 9-year period with no finding of ELF-related pathology or complaints (Kouwenhoven *et al.*, 1967; Singewald *et al.*, 1972). The examination consisted of a medical survey by an internist in consultation with several specialists. Although individual sperm counts varied between normal and abnormal values from year to year, this was considered an artifact. The data did not indicate an electric field influence and the authors' conclusion is that there was no actual effect on sperm counts or sperm motility. In comparison with the Soviet substation workers, the linemen were variably exposed to higher field strengths for shorter periods of time.

In another study of humans, no ELF-related effects were observed on reaction time, blood pressure, pulse rate, electrocardiograph, or electroencephalograph during 3 successive days of testing (Hauf, 1974). Each 3-hour laboratory test session included two 45-minute exposures to a 50 Hz field at 1, 15 or 20 kV/m.

No significant behavioral changes were reported by 20 human subjects exposed to an electric field of 20 kV/m at 50 Hz. The exposed subjects performed as well as the controls in tests of reaction time and in psychological tests (Johansson *et al.*, 1973). The responses to a questionnaire did not disclose a significant difference in perceived discomfort between the test and control groups.

Tests of reaction time, scotopic flicker frequency and hand-eye coordination showed no field sensitivity among persons exposed to a 1 G (45 Hz) magnetic field over a 22.5 hour period (Beischer *et al.*, 1973).

A medical examination of employees exposed to the low level fields (generally less than 0.1 G and less than 0.05 V/m) at the Project Sanguine test facility disclosed no pathology which could be related to the ELF fields. In particular, the neurological symptoms found among substation workers by Asanova and Rakov (1966) were absent (Krumpe and Tockman, 1972; 1974).

Behavioral alterations have been reported in experiments on several species. In man, the daily biological cycle in body temperature, activity, and urinary electrolyte levels (circadian rhythm) was significantly shorter when the subjects were influenced by a small (2.5 V/m), 10 Hz, square wave electric field (Wever, 1967; 1968; 1974). Data from 10 subjects showed an average change of about 1 hour. The 10 Hz frequency was chosen for its similarity to the brain's alpha rhythm and because it is a natural component of the geo-electric field. This careful experiment has not been repeated at higher frequencies or using a sinusoidal wave. The results for exposure to a DC electric field using the same experimental design and apparatus showed no effect, giving greater credibility to the effects demonstrated at 10 Hz.

Human perception of ELF fields in the range of 2 - 12 Hz is indicated by the observation of frequency-dependent changes in reaction times of subjects in an electric field of very low intensity (4 V/m) (Hamer, 1968). The complexity of the experimental design and questions about the data handling indicate the need for reinvestigation. Konig (1971) has also measured changes in human reaction times for low level electric field exposures at 3 and 10 Hz. Perception of low level electric fields at 50 - 60 Hz has not been studied in an experimental situation.

At much higher field strengths, perception of 50 - 60 Hz fields by effects on hair or skin is reported at about 10 kV/m, although the mechanism has not been investigated (Zaffanella, 1972). The perception of magnetic fields of 10 G at 60 Hz has been shown (Schmitt and Tucker, 1973) but the investigators were not certain that all possibility of artifact was excluded. There are also reports of individuals who are able to perceive ELF electric or magnetic fields with uncanny accuracy or sensitivity. For example, Wieske (1963) reports on a few persons who perceived harmonics of the 60 Hz household electric fields as "audible" signals although there was no detectable sound. Schmitt and Tucker (1973) report that some of their subjects are "very certain that they are sensing some kind of field or sensation even though they are not sure whether they hear it, feel it or sense it in some other way." These perceptions are usually correlated with a high degree of accuracy in the test procedure.

The behavioral activities of monkeys exposed to Project Sanguine level fields were unaffected in a number of studies (Grissett and deLorge, 1971; deLorge, 1972; 1973a; 1973b; 1974; deLorge and Marr, 1974). These experiments included an interresponse time (IRT) test using a 9 G magnetic field (45 or 15 Hz) which was negative for changes in the mean values but 2 animals showed significant changes in the distribution of response times during the exposure period. The other studies in this series used combined electric and magnetic fields and were negative with a few exceptions of questionable significance.

ELF electric and magnetic fields at the low levels typical of the Project Sanguine biological effects research program (0.1 to 2 G and 0.06 to 100 V/m at 45, 60 and 75 Hz) were used to examine conditioned behavior in pigeons and rats (Marr *et al.*, 1973). There was no indication that these field conditions could affect conditioned behavior, nor was any preference or avoidance demonstrated in the animal's behavior.

Pilot studies conducted at Project Sanguine level field strengths found no pathological changes during a multi-generational observation of continuously exposed rats (Coate *et al.*, 1970).

Activity levels in mice were affected by a 60 Hz field of about 1 kV/m (Moos, 1964).

The influence of electric transmission lines has been examined in 3 surveys of limited scope. Busby et al.(1974) questioned 18 farmers about the productivity of land crossed by a 765 kV transmission line. The farmers' responses indicate no alterations in the production of corn, wheat, soybeans, beef, or hogs, or in the dairy production of grazing animals, which could be attributed to the line's electrical influence.

In a second survey, 116 of the 125 farmers questioned reported no problems related to the grazing of their animals on land crossed by a 765 kV transmission line (Ware, 1975). Two persons reported that the animals didn't graze under the line, and there were 7 who had experienced shocks. One farmer attributed an animal's death to the lines, but this was not supported by a veterinarian.

A survey of the medical and pharmaceutical services required by the employees of Electricité de France and their families was made over a period of several years (Strumza, 1970). No relation was found between the use of these services and proximity to the line during working and non-working hours.

Effects on Reproduction

The reproductive cycle of slime mold cultures was altered by combined low intensity ELF magnetic and electric fields at 60 or 75 Hz (0.7 V/m, 2 G). The change in the mitotic rate was frequency-dependent, and occurred only after the culture had undergone a long period of exposure (80 - 120 days) encompassing many cell divisions. The phenomenon was reversible after another long period following exposure, and appeared to have a threshold. The relative importance of the electric or magnetic component is not yet known, nor has an explanation for this startling result been proposed.

Strong evidence for mutagenesis in fruit flies was observed (Coate et al., 1970) in a pilot experiment using a 48-hour exposure to combined electric and magnetic fields (20 V/m at 2 G), but two follow-up studies (Bender, 1972; Mittler, 1972) showed no field-related genetic changes. There is no rationale by which production of genetic changes by weak electric and magnetic fields can be explained, a fact which lends additional weight to the two negative studies. As mentioned above, Marino et al. (1974c) observed a large number of chromosomal changes in tumor cells grown in a host animal which was exposed to a strong electrostatic field.

Knickerbocker et al. (1967) found no effects on the fertility of male mice exposed to a very large (160 kV/m) electric field over a 10.5 month period. Although the female progeny (unexposed) of the experimentals and controls grew at the same rate, the male progeny (unexposed) of the experimentals were slightly, but significantly, lower in weight compared with the progeny of control mice. The authors suggest that housing conditions might be responsible for the weight differences, but this does not explain the sex-dependence of the observed effects. No field-related pathology was discovered in autopsies of the exposed mice.

No field-induced mutations were observed in *E. coli* exposed at 45 and 75 Hz to a magnetic field of 1 or 2 G and current which produced electric fields of 10 or 20 V/m in the medium.

It is well known that animals like the newt and salamander can regenerate a limb after amputation, but it is startling that in normally non-regenerating animals, long-term application of a very small DC current can stimulate regeneration. In frogs and rats, a DC current in the nanoampere to microampere range applied to small portions of tissue at the stump stimulates new, differentiated bone, muscle and connective tissues. The extent of regeneration is apparently dependent on placement of the stimulating electrode. The effectiveness of AC currents has not been investigated but the results with DC currents, though not understood, raise the questions of whether environmental AC fields can produce similar current densities and if prolonged exposure to those AC currents could affect the same obscure mechanism of differentiation. The answers take the form of guesses that the largest environmental fields may conceivably produce sufficient current densities, but it is doubtful that evidence for such effects could be developed experimentally.

2.5 EFFECTS OF STRONG MAGNETIC FIELDS

Many effects (for example, on tissue growth and white blood cell formation) have been attributed to inhomogeneous DC magnetic fields of several thousand gauss, and a lesser number of effects to homogeneous DC magnetic fields of several hundred gauss (Barnothy, 1964; 1969).

AC magnetic fields above 100 G (varying at 10 - 100 Hz) evoke a visual response, the "magnetophosphene," probably in the retina, which is not known to be harmful (Barlow *et al*, 1947; Valentinuzzi, 1962). A Soviet survey of industrial workers relates pathological changes in the hands and cardiovascular system which are attributed to magnetic fields of moderate strength (hundreds of gauss at the hands and somewhat less at the trunk) (Vyalov, 1971). Laboratory research in the U.S.S.R. at 200 G (50 Hz) indicates mice and rats have an increased resistance to infection and other experiments indicate that the same field conditions produce an elevation of steroid levels in the blood and an increase in the number of lymphocytes (Vasil'yev *et al*., 1971a; 1971b; Udintsev and Moroz, 1974; Odintsov, 1965). These results suggest that the increased resistance to infection may be a "by-product" of the apparent stress reaction. Another study found pathological changes in many organs of guinea pigs exposed to 50 Hz magnetic fields at 200 G for a period of days or weeks (Toroptsev *et al.*, 1971). Most of the tissue damage appears to be reversible within 30 days. Finally, behavioral changes in rats and an electrophysiological effect in cat brain were reported using a 200 G (50 Hz) magnetic field (Sudakhov and Antimoniy, 1973). In the West, very little research at these field conditions is known, but there is at least one report of an effect on the steroid levels of monkeys exposed at 200 G (DC) (Friedman and Carey, 1972).

Indications that strong magnetic fields may affect biologically important molecules come from several directions. In the first, it has recently been shown that a magnetic field of between 600 and 800 G causes an anomalously large increase in the dielectic constant of lysozyme, a hydrolyzing enzyme (Ahmed *et al.*, 1975). The tentative explanation involves the establishment of a superconducting region in each molecule and a clustering of these regions in the field. If this effect is correct, and has general applicability, there may be consequences for the progress of chemical reactions in strong magnetic fields.

Second, effects on chemical reaction rates in much larger magnetic fields were reported by Haberditzl (1967). A reaction catalyzed by the enzyme glutamate dehydrogenase was slowed by about 10% in a uniform magnetic field of about 50 kG (DC) and by up to 90% in non-uniform fields of similar strength. In a reaction catalyzed by the enzyme catalase, the reaction rate was increased to a marked degree under similar field conditions, but the presence of a paramagnetic species (oxygen) in the reaction confuses the interpretation of the result.

Geacintov *et al.* (1972) used magnetic fields greater than 10 kG to induce orientation of chlorophyll molecules. Entire cells were also oriented by the field, apparently due to diamagnetic anisotropies within the cell.

Finally, at still larger magnetic fields of 100,000 G or more, the DNA molecule can be oriented to a slight degree. This is possible only if the constituent base pairs of the polymer chain are diamagnetically anisotropic. This anisotropy is a reasonable property because of the presence of aromatic rings in the structure of the base pairs. These aromatic rings tend to orient their planes parallel to the external field and hence the DNA chain tends to be perpendicular to the field.

These results indicate that biological molecules and even cells are affected by strong magnetic fields, but it is uncertain that these effects are relevant at the much lower field strengths typical of all but the most exceptional environmental exposures.

The items summarized or referred to above are, in our judgment, the principal information available from experiments on the effects of environmental ELF fields.

2.6 CONCLUSIONS AND DISCUSSION

Although uncertainties are common in all developing fields of knowledge, it is unfortunate that here the uncertainties are compounded by studies of poor quality and public pressure to know "for sure."

For this reason, it is important to place the environmental issue in the perspective of a long history of industrial experience with electric fields which indicates that intermittent exposure to even the strongest such fields has no obvious effect on health or the sense of well-being, even after years of repeated exposure. However, this observation must be tempered by the fact that there has been no full scale epidemiological study of utility workers. The lone 9-year study of 10 linemen (see Sections 2.4, 5.2, and 6.8), which found no health effects, should be expanded to a group large enough to provide statistically meaningful data.

In the Soviet Union, health effects were reported among the employees of a high voltage substation (Sections 2.4, 5.2, and 6.5), where exposures were less intermittent than in the case of the American linemen. The neurological effects reported did not manifest themselves until after a period of several months. The employees' complaints included headaches, fatigue, cardiovascular and digestive disturbances, reduced sexual potency, etc. A medical examination reported changes in blood composition, electrocardiograms, and cardiovascular function.

The U.S. report is not a flat contradiction to the U.S.S.R. study because of important differences in the duration of exposure and the nature of the jobs. Compared to the U.S. linemen, the U.S.S.R. substation workers were exposed for longer periods of time to fields of lower strength and the linemen obtained various degrees of shielding by the use of protective screens, conductive clothing and gloves. Whether these distinctions prove important in understanding the contradictory results remains to be seen.

Persons in fields greater than several kilovolts per meter are subject to spark discharges. Such discharges are annoying and may constitute a safety hazard under special conditions because of their distracting effects. There is also preliminary evidence that these discharges are capable of producing neurological effects, including altered electroencephalograms (EEGs), changes in pulse and respiration rates, and changes in the galvanic skin response (Sections 2.2 and 6.6). In most work areas, the discharge problem can be eliminated or mitigated by the use of electrostatic shields.

The human body behaves as an electrical conductor at the frequencies of interest. For this reason, external fields are greatly attenuated within the body, and only a thin layer of the skin is maximally exposed. Elsewhere within the body, alternating currents flow in a direction parallel to the external field and they flow principally in the electrolyte-rich body fluids. The current paths within the body are unknown, but only a fraction of the total current flows through cells (Sections 2.2 and 4.10).

The possibility of "electric shock" effects can be dismissed, for even the strongest possible external electric fields cannot produce currents in the body strong enough to cause heart fibrillation or tetanic seizures, which are the serious, often fatal, consequences of contact with energized electric conductors.

Humans can consciously perceive 50 - 60 Hz electric fields of about 10 kV/m or more, apparently by erection or vibration of body hairs (Sections 2.4). In the reports of effects on behavioral responses in humans, the field strengths are very much lower (a few volts per meter) and the subject is unaware of the field (Sections 2.4 and 6.1). The mechanism for these effects has not been explained.

Compared with electric fields, the magnetic fields associated with electrical currents may be an environmental factor of equal or greater importance. As with electric fields, a long history of exposures to magnetic fields of 50 - 60 Hz suggests that no obvious risks are associated with acute exposures to fields of moderately great strength (e.g., below 100 G). However, no scientific study among persons regularly exposed supports this observation. The evidence from human experimental investigations to date suggests that there may be biological effects of uncertain significance at fields as low as 1 G (Sections 2.4, 5.2, and 6.2). Unlike the situation with electric fields, the external magnetic field is not attenuated by the body, so that all portions of the body are uniformly exposed to the field. In accord with Faraday's Law, circuitous AC currents are induced in the conducting tissues of a body exposed to a magnetic field.

Investigations of such diverse life forms as birds and bacteria have demonstrated that they respond to the geomagnetic field of 0.5 G (DC) (Section 7.14). Studies of human behavioral responses at low magnetic field strengths are inconclusive, but there is the possibility of perception at levels of about 10 G (Sections 2.4 and 6.1). In man, a distinct visual effect, the "magnetophosphene," is found in stronger AC fields (above 100 G) (Section 7.8).

The conclusion that both electric and magnetic fields of low levels can affect the central nervous system is supported by a number of careful experiments on monkeys and a well-designed study of humans. In monkeys, extensive testing showed an effect on operantly conditioned behavior requiring response within a limited time (Section 2.4). The monkeys were rewarded for lever presses made with a 2.5 second "enable period" which began 5 seconds after the previous lever press. ELF fields in the range of 10 - 100 V/m caused small changes in the inter-response time with the greatest sensitivity at frequencies below 20 Hz, although effects at 75 Hz were also seen. This evidence is not conclusive but is supported by in vitro experiments which demonstrate an interaction between an ELF electric field and the calcium ion equilibrium in brain tissue (Sections 2.4 and 5.6). A similar effect in brain tissues was found using radio frequency waves modulated at several frequencies in the range of 6 - 20 Hz. Similar to the calcium studies, an experiment in brain tissue using a much larger electric field showed changes in the efflux of an important neurochemical, GABA.

Changes in the electrical activity of the brain have been observed utilizing 50 Hz electric fields at 100 kV/m in guinea pi and DC electric fields at 10 kV/m in rats (Sections 2.4 and 7.5), and in experiments in monkeys using both ELF-modulated radio waves and low level ELF electric fields (Section 7.1).

In man, a .1 G (45 Hz) magnetic field appears to have a slight but statistically significant effect on tests of cognitive skill that involved the addition of numbers or memory of symbols (Section 6.3). The research was well planned, executed, and reported, but the conclusions are uncertain. Tests for other behavioral parameters were negative, and the investigators did not conclude that the observed effects were causally related to the ELF magnetic field.

Another class of experiments has investigated the effects of ELF electric or magnetic fields on blood chemistry or blood composition. There is strong evidence from a single, carefully conducted study in men (Sections 2.4 and 5.2) that the serum levels of the triglycerides are elevated a day or two following exposure to a magnetic field of 1 G (45 Hz). Apart from the triglycerides and the pre-beta lipoproteins, no other blood constituents were affected. This alteration of the lipid metabolism is especially notable because of the probable role of elevated lipid levels in atherosclerosis. The authors note that their study is not conclusive because of the small number of subjects and the need for future work in order to determine the relation between the field strength and the lipid levels.

The daily cycle in physiological functions (circadian rhythm) was shortened during exposure to a 10 Hz square wave at a field strength of 2.5 V/m (Sections 2.4 and 6.9). This suggests that external electric sinusoidal fields at 50 or 60 Hz may also alter biological rhythms. However, many questions remai to be answered: the significance of the use of square waves is undetermined; and it is not known if effects would be seen at higher frequencies and amplitudes, if they would persist, or if the natural rhythm would be restored. Finally, if the effect is verified, its significance to human health is not known.

There is a weak data base suggesting effects on the red and white blood cell counts in humans, following exposure to AC electric fields of the order of 10 kV/m (Section 5.2). An investigation of the white blood cells in mice exposed to an electric field of 100 kV/m at 50 Hz supports the general finding of a significant alteration in white cell counts. However, taken across the several studies, the pattern of increases and decreases in each of the different cell types is confused.

A study in rats indicates that changes in blood plasma proteins and the levels of a steroid hormone follow long-term exposure to fields of the order of 15 kV/m. The exposed rats also had lower body weights (Sections 2.4 and 5.3).

In separate investigations, rats exposed to strong electric (15 kV/m) or magnetic (200 G) fields exhibited alterations in steroid hormone levels in the blood, although the experiments have not been confirmed. Such changes are taken as an index of stress, and one of the investigators has advanced the idea that, at sufficient field strengths, electric or magnetic fields may be environmental stressors. Presumably, the stress effects are initiated by action on the nervous system which, in turn, affects the adrenocortical hormones.

DC potentials have a well established role in bone growth and healing (Section 5.10), and direct currents as small as 10^{-8}A, supplied to a small region by electrodes, may cause destructive bone changes (Section 9.1). Pulsed fields also affect bone growth but the effectiveness of sinusoidal AC fields is not well established. Rectification of the AC signals by bone may be possible. In general, current densities in bone or other tissue are not well known but a problem may exist, for example, in the ankle, where the current density can be exceptionally high.

The experimental evidence for mutagenesis by electric fields is contradictory (Section 7.21). Because there are no accepted mechanisms by which such changes could be reasonably expected, the mutagenesis observed in a single pilot experiment is suspect. At very high magnetic field strengths, the occurrence of direct effects on biological molecules makes mutation a theoretical possibility, but at least two studies at very large DC fields (11,000 G and 100,000 G) demonstrated no mutagenesis in fruit flies (Sections 7.21 and 9.5).

Very large magnetic fields (10^5 G) can affect the orientation of polymeric molecules including DNA (Section 7.7). Fields in the range of 10 - 50 kG have orientational effects on certain molecules (e.g., chlorophyll) and cells (e.g., Chlorella) due to their diamagentic properties (Section 2.4) and at much lower field strengths ($\sim$ 600 G) there is recent, preliminary, evidence for effects on cooperative phenomena in biological molecules. These phenomena are important to an understanding of the details of molecular function and structure, and may be more important as research tools than as indicators of an environmentally significant biological effect. Further study would contribute to a better understanding of the importance of the slight structural changes which are known to be important features of molecular interactions in the cell membrane and other systems.

The one firm conclusion that emerges from a review of the existing literature is that relatively weak electric or magnetic fields are capable of evoking neurophysiological or behavioral effects. This is suggested by the behavioral experiments on monkeys, biochemical studies of the calcium efflux in brain tissues, the experience of Soviet substation workers exposed to electric fields, and other industrial workers exposed to magnetic fields and, by inference, by the preliminary evidence for both electric and magnetic effects on the steroid hormones in animals. The serum triglyceride increase could be fit into this pattern, since the lipid metabolism is known to be under hormonal regulation which in turn is regulated by the central nervous system.

The striking conclusions contained in the U.S.S.R. report of health effects among switchyard workers cannot be dismissed out of hand, despite the poor quality of scientific documentation and the imprecision of clinical diagnoses. This situation points out the need for the kind of direct contacts between Soviet and Western scientists which could clarify the issues. It would be helpful if the Soviet findings could be evaluated in the light of an epidemiological study of electric utility workers.

Although the theories proposed to explain the observed effects are relatively undeveloped, we will outline several to indicate the direction of current ideas. In Adey's theory, the apparent sensitivity of neural tissue to low frequency fields is explained in terms of conformational changes in molecules of the cell membrane. This sensitivity is achieved by proposing that the very small initial changes are greatly amplified by triggering other biochemical reactions which can liberate enough energy to affect the synaptic processes of known significance.

In a broader view of biological organisms, Becker proposes that electric and magnetic fields may produce stress-type responses. Stress is accepted as a significant physiological factor related to changes in adrenocortical function; however, the definition of stress and the identification of stressors is difficult. Physiologically, stress elevates the production of the steroid hormones in the adrenal cortex and is so characterist that the presence of elevated blood levels of the steroids is used to indicate stress. In humans, the greatest adrenocortical responses are initiated by certain emotional states involving anxiety and by severe trauma while, for example, extreme exercise (short of exhaustion) does not affect the production of steroids. The question of whether the stress syndrome is physiologically "good" or "bad" does not allow an easy answer and has not yet been thoroughly investigated. In the context of transmission line effects, it would be interesting to investigate the stress response to spark discharges and electric fields separately.

The work done on the electrical aspects of cell differentiation and tissue regeneration involves small DC currents applied for long, continuous periods. Here, only embryonic theories have been suggested and even less can be said of the effectiveness of AC fields in altering tissue growth or of the possibility for pathological tissue changes.

The possibility of mutational or carcinogenic effects appears remote, although there is one report of bone tumors in rats produced by an external electric field. All the evidence shows that strong electric fields do not affect fertility or cause teratogenic effects.

The recent report of altered mitotic rates in slime molds (with no effect on fertility) indicates the possibility for effects on reproductive mechanisms but the interpretation of this experiment is unclear and there is as yet no theoretical basis of understanding for the results.

The consequences of exposure to combined magnetic and electric fields have not been adequately examined, excepting the research at low field strengths conducted for Project Sanguine. It is well known that at much higher field strengths (in electromagnetophoresis) the combined fields have chemical and physical effects not caused by separate exposure to electric or magnetic fields.

Because of existing uncertainties and questions raised by the Soviet reports, it is important that a properly conducted epidemiological study be undertaken. Such a study could answer the question of whether significant effects are occurring among workers in substations, linemen, and workers in other industrial locations where strong electric or magnetic fields occur.

It seems likely that, in practice, the acceptable upper limits for the electric field strength near transmission lines will be determined by the occurrence of spark discharges rather than more subtle effects which are as yet unevaluated. The research conducted to date does not permit unequivocal assessment of the safety of long exposures to ELF fields, but neither is there a basis for alarm concerning exposure to the public. Although the literature indicates physiological changes under certain experimental conditions, the reported effects require verification and then evaluation of their practical significance.

2.7 REFERENCES

Adey, W. R. 1974. The influences of impressed electrical fields at EEG frequencies on brain and behavior. IN: "Behavior and Brain Electrical Activity," Altshuler and Burch, eds. Plenum Press, New York.

Ahmed, N.A.G. et al. 1975. Evidence for collective magnetic effects in an enzyme; likelihood of room temperature superconductive regions. Physics Letters 53A: 129.

Asanova, T.P. and Rakov, A.I. 1966. The state of health of persons working in electric field of outdoor 400 and 500 kV switchyards. Hygiene of Labor and Professional Diseases 5. Translation by G. Knickerbocker available in Special Publ. #10, IEEE Power Engineering Society, Piscataway, N.J. (1975).

Barlow, H.B., Kohn, H.E., and Walsh, E.G. 1947. Visual sensations aroused by magnetic fields. J. Physiol. 148: 372.

Barnes, H.C., McElroy, A.J. and Charkow, J.H. 1967. Rational analysis of electric fields in live line working. IEEE Trans. Pow. App. & Sys. PAS-86, No. 4.

Barnothy, M.F. (Ed.) 1964. "Biological Effects of Magnetic Fields." Volume 1. Plenum Press, New York.

Barnothy, M.F. (Ed.) 1969. "Biological Effects of Magnetic Fields." Volume 2. Plenum Press, New York.

Bassett, C.A.L. and Hermann, I. 1968. The effect of electrostatic fields on macromolecular synthesis by fibroblasts in vitro. (Abstract) J. Cell Biol. 39:9a.

Bawin, S.M. and Adey, W.R. 1975. Effects of weak low frequency electric fields on calcium efflux from isolated chick and cat brain. Abstract of paper presented at Annual Meeting of the U.S National Committee of the International Union of Radio Science (URSI), Boulder, Colo.

Bawin, S.M., Gavalas-Medici, R.J. and Adey, W.R. 1973. Effects of modulated very high frequency fields on specific brain rhythm in cats. Brain Research 58: 365.

Bawin, S.M., Kaczmarek, L.K. and Adey, W.R. 1975. Effects of modulated VHF fields on the central nervous system. Annals N.Y Acad. Sci. 247: 74.

Becker, R.O. 1974b. Prepared testimony before State of New York Public Service Commission, Case 26559.

Beischer, D.E., Grissett, J.D. and Mitchell, R.E. 1973. "Expos of Man to Magnetic Fields Alternating at Extremely Low Frequenc Naval Aerospace Medical Research Laboratory, Pensacola, Fla.. Rept. NAMRL-1180. Available from NTIS as AD 770140 (1973).†

Bender, H. 1972. "System Final Environmental Impact Statement for Research, Development, Test and Evaluation. Naval Electronics Systems Command. PB 199732-F2, C34-C40. Available from NTIS.

Blanchi, D. et al. 1973b. Exposure of mammalians to strong 50 Hz electric fields: 2) effect on heart's and brain's electrical activity. Arch. Fisiol. 70: 33.

Bourgsdorf, V.A. 1972. Discussion of Paper 23.06 at CIGRE, 1972. CIGRE Proc. 1, Group 23, P24 and reprinted in Special Publ. #10, IEEE Power Engineering Society, Piscataway, N.J. (1975).

Busby, K., Driscoll, D. and Washbon, W.E. 1974. "A Field Survey of Farmer Experience with 765 kV Transmission Lines." Agricultural Resources Commission, State of New York, State Campus, Albany, N.Y.

Coate, W.B. et al. 1970. Project Biological Effects Test Program, Pilot Studies, Final Report. Hazleton Laboratories, AD 717408. Available from NTIS.†

deLorge, J. 1972. "Operant Behavior of Rhesus Monkeys in the Presence of Extremely Low Frequency - Low Intensity Magnetic and Electric Fields: Experiment 1." Naval Aerospace Medical Research Lab., Pensacola, Fla. NAMRL-1155. Available from NTIS as AD 754058.†

deLorge, J. 1973a. "Operant Behavior of Rhesus Monkeys in the Presence of Extremely Low Frequency - Low Intensity Magnetic and Electric Fields: Experiment 2." Naval Aerospace Medical Research Lab., Pensacola, Fla. NAMRL-1179. Available from NTIS as AD 764532.†

deLorge, J. 1973 b. "Operant Behavior of Rhesus Monkeys in the Presence of Extremely Low Frequency - Low Intensity Magnetic and Electric Fields: Experiment 3." Naval Aerospace Medical Research Lab., Pensacola, Fla. NAMRL-1196. Available from NTIS as AD 774106.†

deLorge, J. 1974. "A Psychobiological Study of Rhesus Monkeys Exposed to Extremely Low Frequency - Low Intensity Magnetic Fields." Naval Aerospace Medical Research Lab., Pensacola, Fla. NAMRL-1203. Available from NTIS as AD 000078.†

deLorge, J. and Marr, M.J. 1974. "Operant Methods Assessing the Effects of ELF Electromagnetic Fields." IN: "ELF and VLF Electromagnetic Field Effects," M. Persinger, ed. Plenum Press, New York, p. 145.

Deno, D.W. and Zaffanella, L.E. 1975. "Electrostatic Effects of Overhead Transmission Lines and Stations." IN: "Transmission Line Reference Book 345 kV and Above." Electric Power Research Institute, Palo Alto, California.

Friedman, H. and Carey, R.J. 1972. Biomagnetic stressor effects in primates. Physiol. Behav. 9: 171.

Gavalas, R.J. et al. 1970. Effect of low-level, low-frequency electric fields on EEG and behavior in Macaca nemestrina. Brain Research 18: 491.

Gavalas-Medici, R. and Magdaleno, S. 1975. "An Evaluation of Possible Effects of 45 Hz, 60 Hz and 75 Hz Electric Fields on Neurophysiology and Behavior of Monkeys." Univ. of California-Los Angeles, ONR Technical Paper.†

Geacintov, N.E. et al. 1972. Magnetic field induced orientation of photosynthetic systems. Biochim. et Biophys. Acta. 267: 65.

Gibson, R.S. and Moroney, W.F. 1974. "The Effect of Extremely Low Frequency Magnetic Fields on Human Performance: a Preliminary Study." Naval Aerospace Medical Research Laboratory, Pensacola, Fla. NAMRL-1195.†

Grissett, J.D. and deLorge, J. 1971. "Central-Nervous-System Effects as Measured by Reaction Time in Squirrel Monkeys Exposed for Short Periods to Extremely Low-Frequency Magnetic Fields." Naval Aerospace Medical Research Lab., Pensacola, Fla. NAMRL-1137 Available from NTIS as AD 731994.†

Haberditzl, W. 1967. Enzyme activity in high magnetic fields. Nature 213: 72.

Hamer, J.R. 1968. Effects of low level, low frequency electric fields on human reaction time. Comm. Behav. Biol. Pt. A, 2: 217.

Hauf, R. 1974. "Wirkung von 50-Hz-Wechselfeldern auf den Menschen (Effects of 50 Hz alternating fields on man). Sonderdruck aus Elektrotechnische Zeitung-b 26: 381.

IEEE Power Engineering Society, 1975. "Rules and Regulations of Labour Protection at 400, 500, and 750 kV AC Substations and Overhead Lines of Industrial Frequency (in the U.S.S.R.)," SCNTY Orgres, 1971. Translated by G. Knickerbocker in "Study in the U.S.S.R. of Medical Effects of Electric Fields on Electric Power Systems." Special Publ. # 10, Piscataway, N.J.

Johansson, R. et al. 1973. Is there a connection between the electricity in the atmosphere and the function of man? Part 3, 50 Hz field variations. Forvarets Forskningsanstalt Advelning 2: FOA Report C 2627-H5.

Kaczmarek, L.K. and Adey, W.R. 1974. Weak electric gradients change ionic and transmitter fluxes in cortex. Brain Research 66: 537.

Keesey, J.C. and Letcher, F.S. 1970. Human thresholds of electric shock at power transmission frequencies. Arch. Environ. Health 21: 547.

Knickerbocker, G.G., Kouwenhoven, W.B. and Barnes, H.C. 1967. "Exposure of Mice to a Strong AC Electric Field: an Experimental Study." IEEE Trans. Pow. App. Sys. PAS-86, No. 4.

Konig, H. 1971. Biological effects of extremely low frequency electrical phenomena in the atmosphere. J. Interdiscipl. Cycle Res. 2: 317.

Korobkova, V.P. et al. 1972. "Influence of the Electric Field in 500 and 750 kV Switchyards on Maintenance Staff and Means for its Protection." CIGRE 23: 06. International Conference on Large High Tension Electric Systems (28 Aug.-6 Sept., 1972).

Kouwenhoven, W.B. et al. 1967. "Medical Evaluation of Man Working in AC Electric Fields." IEEE Trans. Pow. App. & Sys. PAS-86, No. 4.

Krivova, T.I. 1968. "Electrical Discharge Influence upon a Man." Proc. 3rd All-Union Symp., Hygiene of Labour and Biological Effects of Radiofrequency Electromagnetic Waves, Moscow. Translation by G. Knickerbocker available in Special Publ. #10, IEEE Power Engineering Society, Piscataway, N.J. (1975).

Krivova, T.I., Lukovkin, V.V., and Sazonova, T.E. 1975a. "The Influence of an Electric Field of Commercial Frequency and Discharges on a Human Organism." Abstract by G. Knickerbocker available in Special Publ. #10, IEEE Power Engineering Society, Piscataway, N.J. (1975).

Krumpe, P.E. and Tockman, M.S. 1972. "Evaluation of the Health of Personnel Working near Project Sanguine Beta Test Facility from 1971 to 1972." Naval Medical Research Unit No. 4, Great Lakes, Ill.†

Lott, J.R. and McCain, H. B. 1973. Some effects of continuous and pulsating electric fields on brain wave activity in rats. Int. J. Biometeor. 17: 221.

Marino, A.A. et al., 1974a. "Effect of Continuous Exposure to Extremely Low Frequency Electric Fields on Rats." Veterans Administration Hospital and Upstate Medical Center, Syracuse, N.Y.

Marino, A.A. et al. 1974b. Electric field effects in selected biologic systems. Ann. N.Y. Acad. Sci. 238: 436.

Marino, A.A., Berger, T.J., Austin, B.P., Becker, R.O., and Hart, F.X. 1976a. Evaluation of electrochemical information transfer system. I. Effect of electric fields on living organisms. J. Electrochem. Soc. 123: 1199-1200.

Marino, A.A., Becker, R.O. and Ullrich, B. 1976b. The effects of continuous exposure to low frequency electric fields on three generations of mice: a pilot study. Experientia (Basel) 32: 565-566.

Marino, A.A., Berger, T.J., Austin, B.P., Becker, R.O., and Hart, F.X. 1976c. In vivo bioelectrochemical changes associated with exposure to extremely low frequency electric fields. Bioenergetics and Bioelectrochemistry (in press).

Marr, M.J., Rivers, W.K., and Burns, C.P. 1973. "The Effect of Low Energy, Extremely Low Frequency (ELF) Electromagnetic Radiation on Operant Behavior in the Pigeon and the Rat." Georgia Inst. of Technology, Final Report. Available from NTIS as AD 759415.†

Miller, D.A. 1974. "Electric and Magnetic Fields Produced by Commercial Power Systems." IN: "Biologic and Clinical Effects of Low Frequency Magnetic and Electric Fields," J. Llaurado et al., eds., Charles C. Thomas, Springfield, Ill., p. 62ff.

Mittler, S. 1972. "Low Frequency Electromagnetic Radiation and Genetic Aberrations." Final Report, Northern Illinois Univ. Available from NTIS as AD 749959.†

Moos, W.S. 1964. A preliminary report on the effects of electric fields on mice. Aerospace Med. 35: 374.

Odintsov, Yu. N. 1965. "The Effect of a Magnetic Field on the Natural Resistance of White Mice to Listeria Infection." Voprosy Epidemiologii Mikrobiologii l Immunologii. Available from NTIS in "Translations on Biological Effects of Magnetic Fields," JPRS 62865 (1974).

Polk, C. 1974. "Sources, Propagation, Amplitude and Temporal Variation of Extremely Low Frequency (0-100 Hz) Electromagnetic Fields." IN: "Biologic and Clinical Effects of Low Frequency Magnetic and Electric Fields." Charles C. Thomas, Springfield, Ill

Riesen, W.H. et al. 1971. "A Pilot Study of the Interaction of Extremely Low Frequency Electromagnetic Fields with Brain Organelles." IIT Research Institute, Tech. Memo. No. 3, IITRI Project # 6185.†

Sazonova, T.E. 1967. "Physiological and Hygienic Assessment of Labour Conditions at 400 - 500 kV Outdoor Switchyards." Scientific Pubs., Inst. of Labour Protection of the All-Union Central Council of Trade Unions, Issue 46. Profizdat. Translation by G. Knickerbocker available in Special Publ. #10 IEEE Power Engineering Society, Piscataway, N.J. (1975).

Schmitt, O.H. and Tucker, R.D. 1973. "Human Perception of Moderate Strength Low Frequency Magnetic Fields." IEEE Electromagnetic Symp. Record, June, 1973, pp. 65-70.

Schwan, H.P. 1972. "Biological Hazards from Exposure to ELF Electrical Fields and Potential." Naval Weapons Laboratory Rept. NWL TW-2713.†

Singewald, M.L., Langworthy, O.R. and Kouwenhoven, W.B. 1972. "Medical Follow-up Study of High Voltage Linemen Working in AC Electric Fields. Trans. Paper T 73 154-2, presented at the IEEE Power Engineering Society Winter Meeting, New York, 1973.

Strumza, M.W. 1970. Influence sur la santé humaine de la proximité des conducteur d'électricité a haute tension. (Influence of proximity to high tension electrical lines on human health.) Arch. Mal. Prof. (Paris) 31: 269.

Sudakov, K.V. and Antimoniy, G.D. 1973. "Central Mechanism of Action of Electromagnetic Fields." Uspekhi Fiziologicheskikh Nauk, Russian, No. 2: 101. Available from NTIS as JPRS 60711 (1973).

Toroptsev, I.V. et al. 1971. "Pathologoanatomic Characteristics of Changes in Experimental Animals under the Influence of Magnetic Fields." IN: "Influence of Magnetic Fields on Biological Objects," Y. Kholodov, ed. Available from NTIS as JPRS 63038 (1974).

Udintsev, N.A. and Moroz, V.V. 1974. Response of the pituitary-adrenal system to the action of a variable magnetic field. Bull. Exp. Biol. Med. 77: 641.

Valentinuzzi, M. 1962. Theory of magnetophosphenes. Am. J. Med. Electronics 1: 112.

Vasil'yev, N.V., Boginich, L.F. and Shilyayeva, G.G. 1971a. "Elements of Similarity in the Reaction of the Lymphoid Tissue to Antigenic and Nonantigenic (Alternating Magnetic Field) Stimulation." Moscow Arkhiv Patologii 9: 47. Translation available from NTIS as JPRS 62510 (1974).

Vasil'yev, N.V. et al. 1971b. "Magnetic Fields, Infection and Immunity." IN: "Influence of Magnetic Fields on Biological Objects," Y. Kholodov, ed. Available from NTIS as JPRS 63038 (1974).

Vyalov, A.M. 1971. "Clinico-Hygienic and Experimental Data on the Effects of Magnetic Fields under Industrial Conditions." IN: "Influence of Magnetic Fields on Biological Objects," Y. Kholodov, ed. Available from NTIS as JPRS 63038 (1974).

Ware, B.J. 1975. "Effect of 765 kV Transmission Lines on Animal Grazing Habits." American Electric Power Service Corp., Dec., 1974 (Revised 1975).

Wever, R. 1967. "Ueber die beeinflussung der circadiane periodik des menschen durch schwache elektromagnetische felder (The influence of weak electromagnetic fields on the circadian rhythm in man). Z. F. Vergleichende Physiologie 56: 111. NASA translation X68-13292.

Wever, R. 1968. Einfluss schwacher elektromagnetischer felder auf die circadiane periodik des menschen (Influence of weak electromagnetic fields in the circadian periodicity of man). _Naturwissenschaften_ _55_: 29. NASA translation _X68-18216_.

Wever, R. 1974. "ELF Effects on Human Circadian Rhythms" IN: "ELF and VLF Electromagnetic Field Effects." M. Persinger, ed. Plenum Press, New York.

Wieske, C.W. 1963. Human sensitivity to electric fields. _Biomed. Sci. Instrum._ _1_: 467.

Zaffanella, L.E. 1972. Discussion of "Electrostatic Effects of Overhead Transmission Lines." IEEE Pow. App. and Sys. _91_: 431-432

CHAPTER 3

ENVIRONMENTAL EXPOSURES

Prior to the advent of 20th Century electrical technology, the strongest electric field to which humans were exposed was from the static (DC) atmospheric field, which in fair weather is nominally 120 - 150 V/m (Tromp, 1963). The geomagnetic static field has a maximum value of about 0.5 G at mid-latitudes.

The natural fields vary periodically with both daily and yearly cycles (Polk, 1974). The electric field may increase by nearly 2 orders of magnitude, to 10 kV/m, during a thunderstorm (Dolezalek, 1974). Daily changes in the natural electric fields are caused by the factors that affect electron and ion flow between the earth and atmosphere. In addition to thunderstorms, these factors include the atmospheric concentration of dust and radioactive gases.

Small amplitude (10^{-2} - 10^{-3} V/m) electric fields at low frequency (1 - 10 Hz) are associated with thunderstorm activity and geomagnetic pulsations which induce the telluric fields.* At higher frequencies (above 10 Hz) the natural fields are still weaker. A log-log plot (Figure 3-1) shows a linear relation between electric field strength and frequency, which means

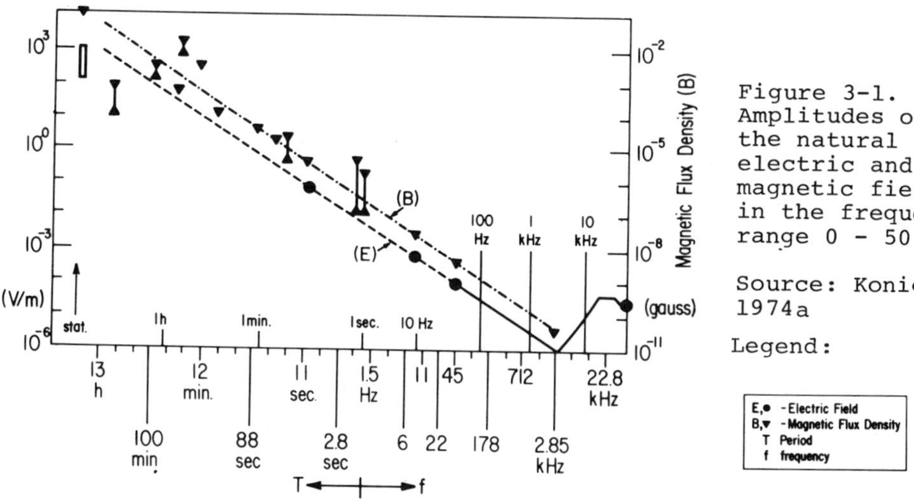

Figure 3-1. Amplitudes of the natural electric and magnetic fields in the frequency range 0 - 50 kHz.

Source: Konig, 1974a

Legend:

---

* Telluric fields are associated with currents flowing in the earth. ("Telluric" is derived from the Latin, "<u>telluris</u>," meaning earth.)

that the natural background field decreases very rapidly with increasing frequency. At 50 - 60 Hz, the electric field natural background is of the order of 10^{-4} V/m (Konig, 1974a). In similar fashion, the natural AC magnetic background falls rapidly to a value of 10^{-8} G at 50 - 60 Hz (Figure 3-1).

Superimposed on the rapidly falling electric field background is a series of resonances (Schumann Resonances) which are produced in the spherical cavity formed between the highly conducting surfaces of the ionosphere and the ground (Figure 3-2). The cavity resonates in a series of up to 5 resonances between 8 and 32 Hz (Polk, 1974).

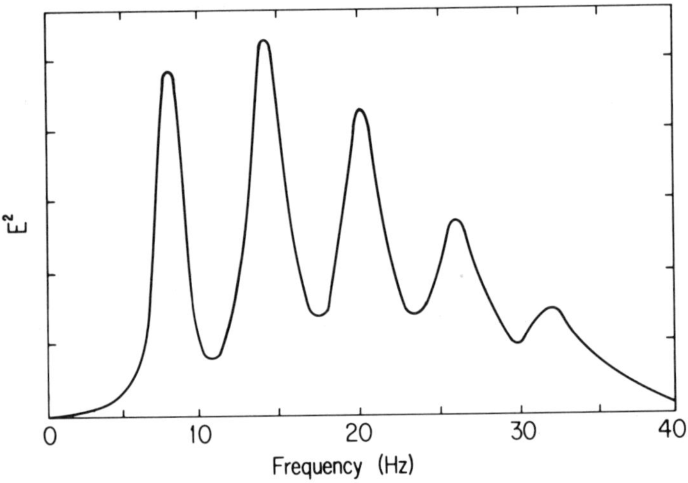

Figure 3-2. Schumann reson-ances in the sp trum of the ver tical electric field (E^2).

Data obtained c July 8, 1972 from 1935 - 19 hours (U.T.).

Source: Polk, 1974 from Keef *et al.*, 1973

The manmade fields are stronger than those of natural origin by many orders of magnitude. Typical 60 Hz fields vary from 3×10^{-3} V/m in a rural area to 10^{-1} V/m outside an electrified building to 10^{1} to 10^{2} V/m near electric appliances (Table 3-1). Typical values for ambient 60 Hz magnetic fields range from about 10^{-3} to about 10 G, depending on proximity to appliances (D. A. Miller, 1974) (Table 3-2). Higher electric fields exist near high voltage transmission lines, substations, and certain industrial equipment. Stronger magnetic fields occur near large magnets, electrical furnaces or other equipment using high currents. The potential use of cryogenic or superconducting transmission facilities would introduce another localized source of strong magnetic fields.

The distribution of the electric fields about overhead transmission facilities has been well described by both calculations and field measurements. The several possible geometric arrangements of the conductors (vertical or horizontal, 1 circuit or 2), and details of design such as the diameter and configurati

TABLE 3-1. ENVIRONMENTAL 60 Hz ELECTRIC FIELDS

a) Electric Field (v/m) measurements outside the home*
-- in a rural area 0.003 ± 0.006 (S.D.), (N = 202)
maximum field, 0.04

-- near home grounds 0.094 ± 0.088
maximum field 0.525

b) 60 Hz Electric Fields at the Center of Various Rooms in a Typical Home**

Room	Electric Field (volt/meter)
Laundry Room	0.8 v/m
Dining Room	0.9
Bathroom	1.2
Bathroom	1.5
Bedroom	2.4
Kitchen	2.6
Living Room	3.3
Bedroom	5.5
Bedroom	7.8
Hallway	13.0

c) 60 Hz Electric Fields Measured 30 cm from Various Electrical Appliances**

Incandescent Light Bulb	2 v/m
Electric Range	4
Clock	15
Vacuum Cleaner	16
Coffee Pot	30
Color TV	30
Hair Dryer	40
Vaporizer	40
Toaster	40
Phonograph	40
Hand Mixer	50
Iron	60
Refrigerator	60
Stereo	90
Broiler	130
Electric Blanket	250

---

* D.A. Miller, 1974, derived from Hartell et al., 1972
** D.A. Miller, 1974, derived from Valentino, 1972

TABLE 3-2. ENVIRONMENTAL 60 Hz MAGNETIC FLUX DENSITIES*

a) Typical 60 Hz Ambient Magnetic Flux Densities

Location	Flux Density
University of Pennsylvania Hospital	0.002-0.004
Princeton Hospital	0.0003-0.001
Park Falls (Wisc.) Hospital	0.0005-0.008
Industrial Plant, Park Falls (Wisc.)	0.003-0.06
Scientific Laboratory, Pensacola, Florida	0.005-0.01
IIT Research Institute Office Areas	0.05
Private Dwelling	0.001-0.101

b) Localized 60 Hz Magnetic Flux Densities Produced by Some Electrical Appliances

<u>10-25 Gauss</u>
325 Watt Soldering Gun
Magnetic Stirrer
Power Feeder Cable
Hair Dryer

<u>5-10 Gauss</u>
Can Opener
140 Watt Soldering Gun
Fluorescent Desk Lamp
Kitchen Range
Electric Shaver

<u>1-5 Gauss</u>
Bench Grinder
Arc Welder
Food Mixer
Power Transformer
Induction Motor
Color Television Set
Food Blender
Electric Drill
Portable Heater

<u>0.1-1.0 Gauss</u>
Toy Auto Transformer
Garbage Disposal
Clothes Dryer
Black/White Television Set
Vacuum Cleaner
Heating Pad
Electric Toaster
Bell Transformer

<u>0.01-0.1 Gauss</u>
Home Electric Service Unit
Kitchen Fluorescent Lamp
Dishwasher
Laundry Washer
Phonograph
Calculator
Electric Iron

<u>0.001-0.01 Gauss</u>
Refrigerator

---

* D.A. Miller, 1974, derived from Hartell *et al.*, 1972

of conductor bundles, and height of the wires above ground, all influence field strengths near the ground. The most significant parameter is the height of the conductor above ground. Typical values of the maximum electric field near overhead transmission lines range from 7.5 kV/m for 345 kV lines* (double circuit) (Westinghouse, 1974) to 15 kV/m for the 1100 kV lines (Deno and Zaffanella, 1975). These values are the "ground level field strengths." The vertical field near the ground changes slowly so that the field at the height of a man (1.8 meters) would not be more than 10% higher than the ground level value.** Typical graphs of field strengths vs. horizontal distance from the center line are shown in Figure 3-3. At horizontal distances greater than approximately twice the line height, the electric field decreases inversely with distance. Figure 3-3b shows that field cancellation which may be arranged by appropriate phasing of adjacent circuits. In Figure 3-3c, the <u>peak</u> electric field under a 1050 kV line, compared to a 525 kV line, is not increased by as much as the ratio in voltage, due to the increase in conductor height. But, and for the same reason, the fields at various distances (e.g., 30 meters) from the center line may be several times the voltage ratio. These figures also indicate good agreement which can be obtained between measurement and calculation of the field. Typically, the right of way purchased for a transmission line is from 35 to 70 meters, so that at the edge of the right of way electric fields are of the order of 10^3 V/m. The electric fields associated with transmission lines are the strongest electric fields to which persons, animals, and plants may be routinely exposed.

The magnetic fields associated with the overhead transmission lines are usually of the order of a few tenths of a gauss at maximum and so far have attracted less attention. The magnetic field also decreases inversely with distance far from the line, so that fields at the edge of the right of way may be about 10^{-1}-10^{-2} G.†

---

*The phase-to-phase voltage at a nominal operating voltage is used to rate transmission lines. The phase-to-ground voltage is 1 $\sqrt{3}$ times as large.

**Detailed data and graphs are available (Westinghouse, 1974; Bracken, 1976; C. Shih, personal communication, 1974; Deno and Zaffanella, 1975; Barthold *et al*., 1971a; Schneider *et al*., 1974 and Lyskov *et al*., 1975).

†The magnetic flux density, B, about a single long wire is given by Ampère's circuital relation,

$$B = \frac{\mu_0 I}{2\pi r}$$

where I is the current in the wire, r the radial distance from the wire, and μ_0 is the permittivity of air (taken equal to the free space value, $4\pi \cdot 10^{-7}$ henry/m). Consequently, at 2 meters from a wire carrying 1000 amperes the magnetic flux density is 1×10^{-4} tesla = 1 gauss. At 25 meters the field is 8×10^{-6} tesla = 0.08 G.

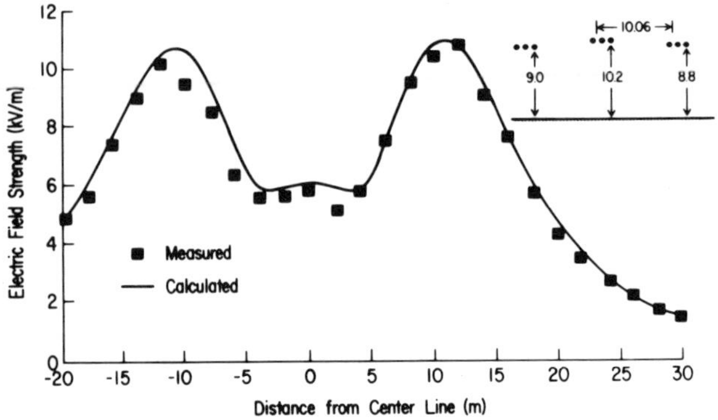

Figure 3-3a. Compute and measured field strength near 525 kV line at 1 meter heig above ground. The conductor heights above ground, indi- cated in the sketch, are minimal and thus give maximal field strengths (Bracken, 1976.

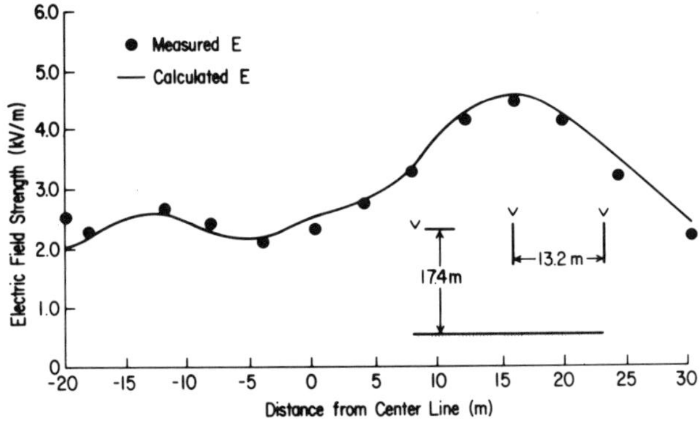

Figure 3-3b. Field strengths at 1 meter height near a 525 kV line adjacent to another 525 kV line centered at -38.1 me ter. The conductor heights, shown in th inset, are about twi the height of those in Fig. 3-3a (Bracke 1976).

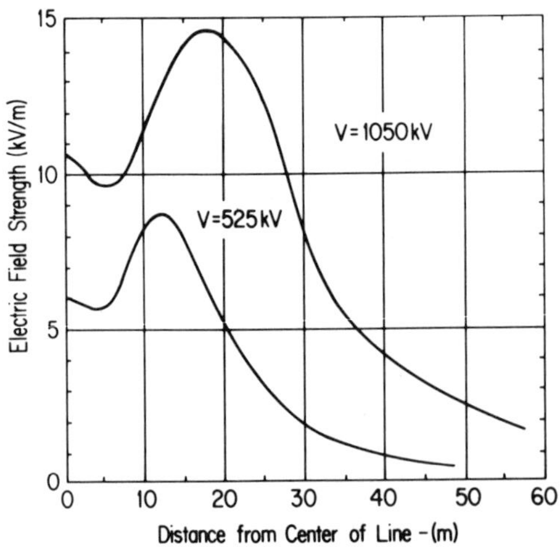

Figure 3-3c. Comput ground level electri field strengths near 525 and 1050 kV line (Deno and Zaffanella 1975).

The advance to higher voltages requires the use of conductors of larger effective diameter to reduce corona losses and the concurrent ozone production, audible noise, and radio and TV interference. This is achieved by designing bundles of wires spaced apart by about 50 cm coincidentally increasing the current-carrying capabilities of the line. However, this manyfold increase in current-carrying capability is not fully utilized in present designs, so that the currents and resulting magnetic fields associated with the higher voltage transmission lines are not very much different from those at lower voltages. In all cases, the magnetic field varies with the actual current flow which is in response to the line load.

Individual exposures to fields depend upon the circumstances and habits surrounding the use of electrical appliances at home or in the job situation. Utility employees working in substations or the maintenance of transmission lines form a special group exposed continually to large fields and persons living close to high voltage transmission lines may also be exposed more than the majority of the population. A hunter using the cleared right of way as a passageway or for his hunting poses a special instance of exposure. The use of the transmission line corridor for recreational purposes has been proposed (Randall, 1973). This would create another category of exposed persons.

Environmental exposures are often categorized as acute or chronic. In this case, there are no available guidelines, but it seems reasonable that persons living in electrified homes are chronically exposed to fields of at least a few volts per meter, and persons spending much time near high voltage equipment or working for power companies maintaining such equipment may be acutely exposed for several days a week to fields of several thousand volts per meter. Similarly, magnetic field exposures for most of the population are chronically in the range of less than 1 G at 60 Hz and only persons working near high current machines would be exposed to levels exceeding a few gauss over a substantial period of the workday. The further quantification of exposures is necessary.

3.1 REFERENCES

Barthold, L.O. *et al*. 1971a. "Electrostatic Effects of Overhead Transmission Lines" Part I. "Hazards and Effects." *IEEE Paper 71 TP 644-PWR*. Transmission and Distribution Committee of the IEEE Power Engineering Society (1971).

Bracken, T.D. 1976. Field measurements and calculations of electrostatic effects of overhead transmission lines. *IEEE Trans. Pow. App. and Sys., PAS-95*: 494-504.

Deno, D.W. and Zaffanella, L.E. 1975. "Electrostatic Effects of Overhead Transmission Lines and Stations." IN: "Transmission Line Reference Book 345 kV and Above." Electric Power Research Inst., Palo Alto, Calif.

Dolezalek, H. 1974. "Atmospheric Electricity." IN: "Handbook of Chemistry & Physics," 55th ed. CRC Press, Cleveland, Ohio, p. F-196.

Hartell, W.K., Miller, D.A., and Abromavage, M.M. 1972. "Project Sanguine's Environmental Compatibility Assurance Program." *IEEE Int. Electromagnetic Compatibility Symposium Proc.*, p. 125.

Keefe, T.J., Etzold, H. and Polk, C. 1973. "Detection and Processing of ELF (3-30 Hz) Natural Electromagnetic Noise." Univ. of Rhode Island, Kingston, R.I. *Report AFCRL-TR-73-0077*.

Konig, H.L. 1974a. "ELF and VLF Signal Properties: Physical Characteristics." IN: "ELF and VLF Electromagnetic Field Effects," M. Persinger, ed. Plenum Press, New York, p. 14.

Lyskov, Y.I., Emma, Y.S. and Stolyarov, M.D. 1975. "The Factors of Electrical Field That Have an Influence on a Human." IN: "Three Russian Papers on EHV/UHV Transmission Line and Substation Design." Uhl, Hall and Rich, Div. of Charles T. Main, Inc., Boston, Mass.

Miller, D.A. 1974. "Electric and Magnetic Fields Produced by Commercial Power Systems." IN: "Biologic and Clinical Effects of Low Frequency Magnetic and Electric Fields." Llaurado *et al*., eds., Charles C. Thomas, Springfield, Ill., p. 62ff.

Polk, C. 1974."Sources, Propagation, Amplitude and Temporal Variation of Extremely Low Frequency (0-100 Hz) electromagnetic fields." IN: "Biologic and Clinical Effects of Low Frequency Magnetic and Electric Fields." Llaurado *et al*, eds. Charles C. Thomas, Springfield, Ill.

Randall, E.E. 1973. "Multiple Use Potential along Power Transmission Rights of Way in Power Lines and the Environment." R. Goodland, ed. The Cary Arboretum of the N.Y. Botanical Gardens, Millbrook, N.Y., p. 101ff.

Schneider, K.H. et al. 1974. "Displacement Currents to the Human Body Caused by the Dielectric Field under Overhead Lines." CIGRE Rept. #36-04, presented at Internat'l Conf. on Large High Voltage Electric Systems, 112 Blvd. Haussmann--74008 Paris.

Tromp, S.W. 1963. "Medical Biometeorology." Elsevier Publ. Co., Inc., New York, p. 74ff.

Valentino, A.R. 1972. "A Small ELF Electric Field Probe." Proc. IEEE Internat'l Electromagnetic Compatibility Symp., p. 265.

Westinghouse Electric Corp., 1974. "Research of Corona and Electric Field Effects of EHV Transmission in Eastern Long Island." Report #AST 74-951, Westinghouse Power Systems Planning Advanced Systems Technology, p. 205.

CHAPTER 4

BIOPHYSICAL PRINCIPLES AND A BIOPHYSICAL DESCRIPTION OF EXPOSURE TO ELF FIELDS

The physical changes that occur when an object is placed in an electric or magnetic field can be measured by the internal electric field, internal magnetic field, current density, or total current. This section reviews the interrelationships of the essential physical parameters such as field strengths, currents, capacitances, and conductivities.

We begin by discussing the reasons for considering the electric and magnetic fields separately in the quasi-static approximation.

4.1 THE QUASI-STATIC APPROXIMATION

The extremely long wave lengths characteristic of ELF frequencies -- about 5,000 km (3100 mi) at 60 Hz -- make it appropriate to discuss the uncoupled electric and magnetic fields and disregard any questions of radiation. This is so because a consideration of an overhead transmission line as if it were an antenna shows that one is always in the near or static zone where E, the electric field, and H, the magnetic field intensity, do not have a simple phase or spatial relation. The energy in the fields is not radiated away into space, but collapses back onto the line, so that power can be transmitted rather than radiated. A power line is an exceedingly poor transmitting antenna at 60 Hz.

The term "static field" is used to describe the near field of a dipole antenna because the spatial properties of the electric near field are the same as the field of a static electric dipole (Stratton, 1941). In our case, this use of the term also implies that we are not in the radiation zone. In addition, we note that the slowly varying field affects objects in a way that often allows us to think of the similar effects in a static field. Hence, the terms "quasi-static field" and "electrostatic induction" are found in the ELF literature. Electrostatic induction (the charging of a conducting body placed in an electric field) is distinct from electromagnetic induction (Faraday's Law) in which an electric field results from a changing magnetic flux (as in a transformer).

4.2 ENERGY DENSITY IN THE ELECTRIC FIELD

The energy density represented by the electric field is very small. For example, a 10 kV/m electric field contains an energy density of 4.42×10^{-8} joules/m^3. In a unit more appropriate to a biological cell, the energy density of a 10 kV/m electric field is about 10^{-26} calories per cubic micron, or 2.8×10^{-7} electron volts per cubic micron. The energy represented by the field is deposited in body tissues by ohmic heating of the tissues through which the current flows and, in comparison, the dielectric losses are negligible (see Section 4.4). The power density absorbed by a man in a 10 kV/m field is about 10 μwatts/m^3 compared with a metabolic rate of heat production 10^8 times greater. Thus, the energy density remains negligible even at the highest field strengths encountered in practice and, as shown later (Section 4.8), we need not be concerned with the heating of cells or tissues.

4.3 DIELECTRIC CONSTANT

The dielectric properties of biological materials have been investigated over a wide range of frequencies. The extreme variations in the magnitude of the dielectric constant and the lesser variations in the conductivity have been categorized according to frequency as the α, β, and γ dispersion regions (see Figure 4-1). Only the α-region is of importance at low frequencies. The β-dispersion commonly occurs at 1 - 500 MHz and the γ-dispersion, characterized by the relaxation time of the water molecule, occurs at about 3 GHz.

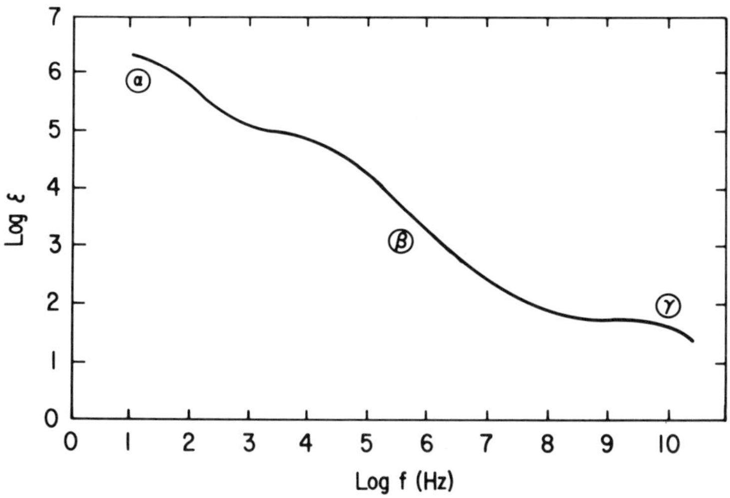

Figure 4-1. The dielectric constant of muscular tissue as a function of frequency decreases in 3 major steps from about 3×10^6 to about 30 as the frequency increases from 10 Hz to 20,000 MHz. The 3 steps are identified as the α-, β-, and γ-dispersions.

Source: Schwan, 1957

The dielectric constant of tissues decreases rapidly as the frequency increases, even over the range of 10 - 100 Hz, because of the properties of the cell membrane (Schwan, 1957; Schwan and Kay, 1957a; Hasted, 1973). (See Table 4-1.) At low frequencies the cell membrane has the properties of a capacitator with a low loss dielectric of enormous dielectric constant. The static and low frequency

TABLE 4-1. CONDUCTIVITIES AND DIELECTRIC CONSTANTS OF DOG TISSUES *IN SITU**

a) Dielectric Constants in Units of $\varepsilon_o = 8.85 \times 10^{-12}$ F/m

	10 Hz	100 Hz	1 kHz	10 kHz
Lung	$8. \times 10^6$	4.5×10^5	0.9×10^5	$3. \times 10^4$
Muscle	10.	8.0	1.3	6.0
Liver	16.	8.5	1.5	6.0
Heart muscle	7.	8.2	3.2	10.
Fatty tissue	-	1.5	0.5	2.

b) Conductivities, σ, in mho-m^{-1}

	10 Hz	100 Hz	1 kHz	10 kHz
Lung	.090	.091	.096	.105
Muscle	.104	.114	.120	.132
Liver	.119	.125	.130	.146
Heart muscle	.104	.108	.118	.167
Fatty tissue	-	-	.02 - .066	-

*Schwan, 1957. The data average about 20 samples each. Standard deviation about 20%. However, the variation with frequency is significant to about 1% for the values. No good *in situ* data for the dielectric constant are available below 100 Hz.

dielectric properties of cell suspensions (as for the red blood cells in the serum) and tissues have been investigated experimentally and theoretically (Cole and Cole, 1941; Schwarz, 1962; Schwarz *et al.*, 1965; Schwan *et al.*, 1962; Heller *et al.*, 1963; Schwan, 1963; Schwartz, 1973).

The dielectric properties of protein molecules in solution have also been studied. The effect of an external static field on the ions and protons within these large molecules produces an effective electric dipole moment and a force between the molecules (Kirkwood and Shumaker, 1952a; 1952b).

4.4 CONDUCTIVITY

The resistive properties of tissues are far less dispersive than their dielectric properties. As shown in Table 4.1, the conductivity falls relatively slowly below 10 Hz. The overall property of the body interior is that of a conductor with a resistivity of 5 - 10 ohm-meters. In the body, the displacement current due to polarization is very much less than the current carried by the free charges (ions) despite the high dielectric constant of tissues. The rich ionic environment is reflected in the fact that, excepting bone and fat, body tissues are about 75% water in which electrolytes are dissolved. Barnes *et al.* (1967) calculated the ratio of the free charge to the polarization charge. They derive the relation,

$$\frac{\text{free charge}}{\text{polarization charge}} = \frac{1}{\omega\rho\,(\varepsilon-\varepsilon_o)}$$

where ω is the frequency in radians; ρ, the materials resistivity; ε, the material's dielectric constant; and ε_o, the dielectric constant of free space, which is 8.85×10^{-12} farads per meter (F/m). Assuming a dielectric constant of 80 and a conductivity of 1 ohm-meter the ratio is several million, but by assuming a more realistic value of 10^6 for the relative dielectric constant in body tissue, and 8 ohm-meter for the conductivity (Schwan, 1972), the ratio of free to bound charges is about 40, allowing the same conclusion as reached by Barnes: the free charge conductivity predominates within the body tissues. This also means that the displacement current associated with capacitance can usually be ignored. See also Schwan and Kay (1957a; 1957b).

4.5 MODELS OF A CONDUCTOR IN AN ELECTRIC FIELD

Because of the predominance of free charges, the electrical effects of placing a body in a static or low frequency electric field can be examined by analogy with the simpler problem of placing a conducting sphere in an electric field. The conducting sphere in a static (DC) field is often a textbook problem (Stratton, 1941), in which it is shown that the conducting sphere becomes polarized by the formation of surface charges and acts like a dipole. Within the conducting material, the electric field, of course, is zero and no current flows.

However, if the same conducting sphere is placed in an alternating electric field, a current will flow as the sphere polarizes first in one direction, and then the other. The flow of current within the conducting sphere is determined by the internal field and the geometry of the conductor, while the resultant internal electric field is determined by the internal conductivity (Ohm's Law). The total current which passes through the sphere is calculated by considering the isolated sphere as a capacitor in which the two hemispheres are at different potentials as determined by the boundary conditions. The capacitance is given by the ratio of the charge Q to the potential V,

$$C = Q/V.$$

The current through the sphere, I, is,

$$I = \frac{dQ}{dt}$$

Thus, if V is a function of time with frequency ω,

$$V(t) = V_o \cos(\omega t)$$

then

$$I(t) = \omega C V_o \sin(\omega t). \quad (1$$

The interior currents which flow move charge to the surface of the sphere so that the boundary conditions are once again met.

This model has been applied to the case of a man (see Figure 2-2) who is isolated from ground in a uniform electric field. Utilizing prolate spheroidal coordinates (Barnes <u>et al</u>., 1967), the calculation proceeds by solving the boundary value problem for Laplace's equation. The electric field is readily calculated from the potential found as the solution to Laplace's equation, and the current density, $\vec{J}$, follows from the internal electric field, $\vec{E}_{int}$ and the conductivity, σ,

$$\vec{J} = \sigma \vec{E}_{int} \text{ (Ohm's Law)}.$$

The results show that for any direction of the applied field, the current density inside the body is uniform and parallel to the original electric field. Barnes <u>et al</u>. (1967) have tabulated the results of this calculation to allow ready determination of the current density and total current for a prolate spheroid in an external field parallel to either the major or minor axis.

Using these tables, one finds that, for a man approximated as a prolate spheroid of major axis 1.8 meters and minor axis 0.4 meter, in a uniform electric field of 10 kV/m, the current density flowing in the body is 520 $\mu A/m^2$ and the total current i 65 μA. Similar calculations (see Table 4-2) show that in a 10 kV/m field, a rat has a current density of 75 $\mu A/m^2$ and a

TABLE 4-2. CURRENTS AND CURRENT DENSITIES CALCULATED FOR MAN, RAT AND MOUSE IN A 10 kV/m (60 Hz) ELECTRIC FIELD *

	Parameters for Model (m)			Electric Field Strength (kV/m)	Total Induced Current (μA)	Current Density (μA/m^2)
	a	b	c			
Man —	0.9	0.2	0.875	10	65.	520
Rat —	0.075	0.025	0.071	10	0.45	75
Mouse —	0.04	0.015	0.037	10	0.14	76

*The data are derived from the tables in Barnes et al. (1967) using the indicated parameters for a, the semi-major axis, b, the semi-minor axis, and c, half the distance between the foci.

total current of 0.45 μA, and the mouse a current density of 76 $\mu A/m^2$ and a total current of 0.14 μA. Although the dielectric constant and conductivity assumed by Barnes et al. (1967) are unrealistic (Schwan, 1972), the results for the current density are not affected. This is because the current density is independent of the conductivity of the body so long as the conductivity is very much greater than the complex conductivity of air. Or, one may note intuitively that the internal electric field, E_{int}, depends inversely on the conductivity, σ, so that the current density, $\sigma \times E_{int}$, is independent of the conductivity. Of course, since the model assumes a conductor the relative dielectric constant of the body is not considered (Schwan, 1972).

The situation is slightly different if we place the conducting body close to ground, since now we can characterize the situation by adding an equivalent parallel capacitance formed by the body as one electrode and the ground as the other (Figure 2-1). Thus, the spacing between the body and ground becomes critical in determining the capacitance and the resulting current through the conducting electrode-body. If, as often occurs, a person standing on the ground has considerable leakage resistance to the ground, then we can represent the man standing on the ground by the equivalent circuit of Figure 2-1c. Measurements of the body current in this situation have been made (Deno and Zaffanella, 1975) and indicate that the capacitance of the human body is approximately 100 pF.

Using the data from Deno and Zaffanella (1975), which is reproduced as Table 4-3, we get a current of 200 μA for a man standing in an unperturbed electric field of 10 kV/m. If the effective height of a man above ground is chosen to be about one-quarter of his total height, or a half-meter, then the man's open circuit potential is

$$V_{OC} = h \times E, \tag{2}$$

where h is the effective height above ground. The capacitance is related simply to the current and open circuit potential V_{OC} by

$$C = I/(\omega V_{OC}), \tag{3}$$

from which we calculate a capacitance of just over 100 pF using the assumed effective height of 0.5 meters. This capacitance, which is in agreement with the measurements, is reasonable.

As the man is removed further and further from the ground plane, his capacitance to the ground decreases until his effective capacitance as an isolated conductor dominates. This result is seen in data showing that the capacitance measured for persons standing above the earth on an insulated platform approaches 50 pF, at heights of more than a few meters (Deno and Zaffanella, 1975). If we calculate an effective capacitance for the isolated man conducting 65 μA of current in an electric field of 10 kV/m

TABLE 4.3. ELECTROSTATICALLY INDUCED CURRENTS AND CAPACITANCES IN MAN, HORSE AND COW*

Subject		Induced Current (A per V/m)	Capacitance (Farads)
People:	1.8 m, 105 kg	$0.020 \cdot 10^{-6}$	$110 \cdot 10^{-12}$
	1.75 m, 68 kg	$0.018 \cdot 10^{-6}$	$100 \cdot 10^{-12}$
Horse:	Length 2.03 m, shoulder height 1.26 m Belly clearance 0.6 m, 385 kg	$0.027 \cdot 10^{-6}$	$180 \cdot 10^{-12}$
Cow:	Length 2.03 m, shoulder height 1.17 m Belly clearance 0.47 m, 318 kg	$0.024 \cdot 10^{-6}$	$200 \cdot 10^{-12}$

Measurement on 5 cm of insulation

*Deno and Zaffanella, 1975

(see above) and use 0.5 for the effective separation of the capacitor electrodes we find an effective capacitance of about 40 pF, in good agreement with the measurements. Finally, using a spherical model, Schwan (1972) has calculated that 100 pF is the upper limit for man's capacity to free space. It is not surprising that this spherical model gives about twice the capacitance measured by Deno and Zaffanella.

In the Russian literature (Asanova and Rakov, 1966; Korobkova et al., 1972), potentials and not electric fields have been cited, so that it is necessary to convert the potential information into electric field strengths, which can be done by assuming the capacitance of 100 pF justified above. For example, Asanova and Rakov (1966) cite a body current of 230 μA at a potential of 26 kV. For a capacitance of 100 pF at 50 Hz ($\omega = 314\ s^{-1}$), we find the open circuit potential (V_{OC}) is 7.3 kV. If we now assume that the 26 kV potential is located at the height of a man, there is a potential gradient in free space of 26 kV over the 1.8 meter height of a man. Thus, the electric field is 26 kV/1.8 meters, or 14.45 kV/m. Using the effective height of 0.5 meter found above, the current through the man (Equations 2 and 3) is 230 μA, in agreement with their data. Thus for the 100 capacitance whose effective height is 0.5 meter, the 26 kV potential at 1.8 meters results in open circuit potential of 7.3 kV and a current of 230 μA. The Russian potential measurements may be interpreted as electric fields simply by dividing the potential (in kV) by 1.8 meters to arrive at an electric field strength in kV/m.

4.6 ELECTRIC FIELD DISTORTION

The electric field near the surface of a conducting spheroid (Figure 4-2) is distorted so that the field on the surface is not equal in strength to the initial field. Simarily, a person distorts the electric field nearby so that the actual field at the surface of the head, arm or knee may be stronger than the original field (Figures 2-2 and 4-2b). The degree of distortion and intensification depends on the region's curvature and position and may be gauged by calculations made for spheroids of varying shape oriented parallel or perpendicular to the field. Figure 4-3 shows that for a prolate spheroid whose major and minor axes are in the ratio 4.5 to 1 (approximating the human body) the field at the top is increased by a factor of 15 and for a ratio of 10 (approximating the arm) the intensification is 50-fold.

Actual measurements of the surface fields were made on linemen positioned very near a live transmission line conductor and these data (Table 4-4) indicate the largest fields are at the head. Similar data are not available for a person standing on the ground under a transmission line or in another region of uniform fields. The analysis of models in a uniform vertical field (e.g., Schneider et al., 1974) indicates the greatest surface fields are at the head or an outstretched fingertip.

TABLE 4-4. FIELD DISTORTION BY HUMAN BODY *

Body Location	Local Electric Field Strength (kV/m)
Top of head	430
Back of head	350 (calculated)
Face	300
Shoulder	215 - 275
Chest	98.5
Back	200 - 275
Groin	20
Knee	100 - 150

* The data, measured on a lineman close to an energized 200 kV conductor, illustrate the variability of the fields at the body surface. The conductor is approximately 20 cm from the lineman's chest and he is standing on a metal platform surrounded by a 5 cm band at about waist height. Data adapted from Barnes et al. (1967).

Figure 4-2. Field distortion by conductors

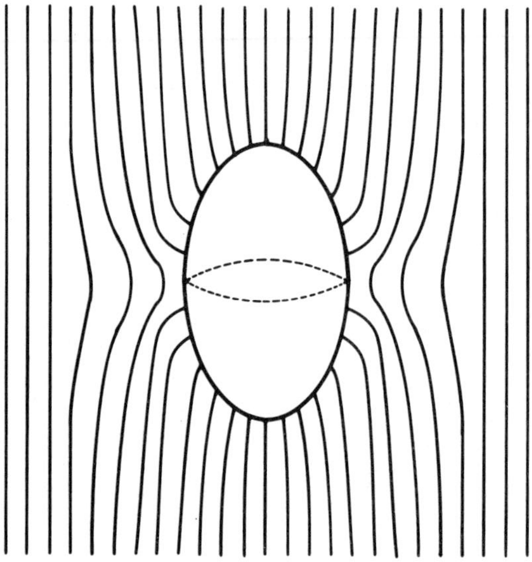

Figure 4-2a.

Sketch of electric field distortion by a prolate spheroidal conductor placed in a uniform field parallel to the major axis.

Figure 4-2b.

Sketch of distortion in a uniform electric field by a grounded man.

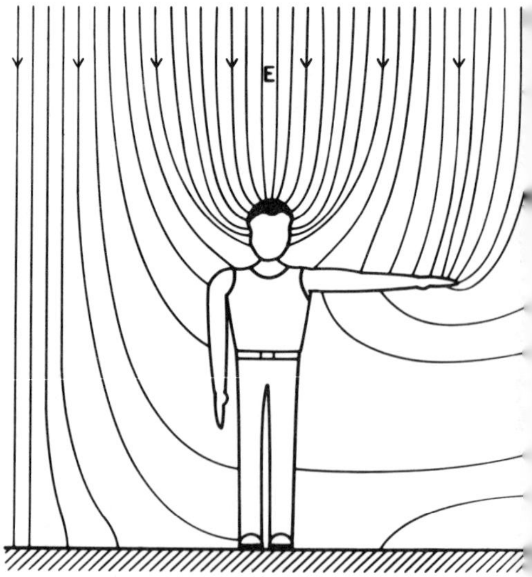

Figure 4-3. Electric field on the surface of a prolate spheroid
Source: Barnes *et al.*, 1967

(A)

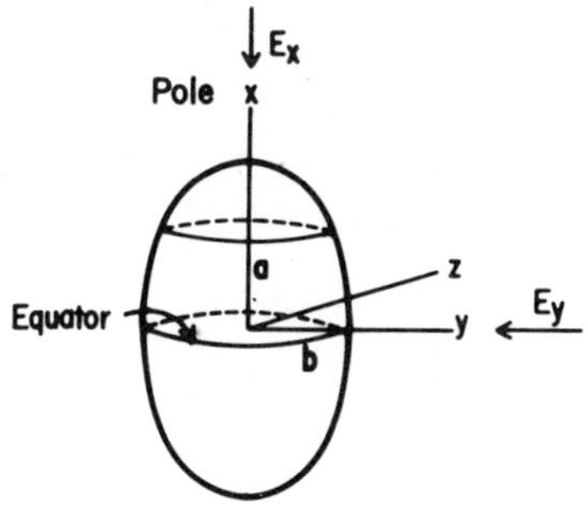

Figure 4-3a. Prolate spheroid aligned with major axis, of length "a", parallel to the applied field, E_x or minor axis "b" parallel to E_y.

(B)

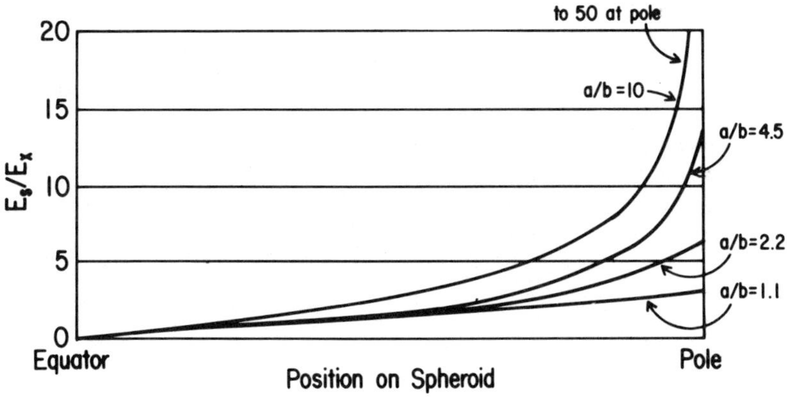

Figure 4-3b. Electric field on surface of the prolate spheroid, E_s, relative to the applied field, E_x, as a function of position for several spheroids. "a/b" is the ratio of the major and minor axes; a/b = 4.5 approximates the human body and a/b = 10 an arm.

Figure 4-3. Electric field on the surface of a prolate spheroid (Cont'd)

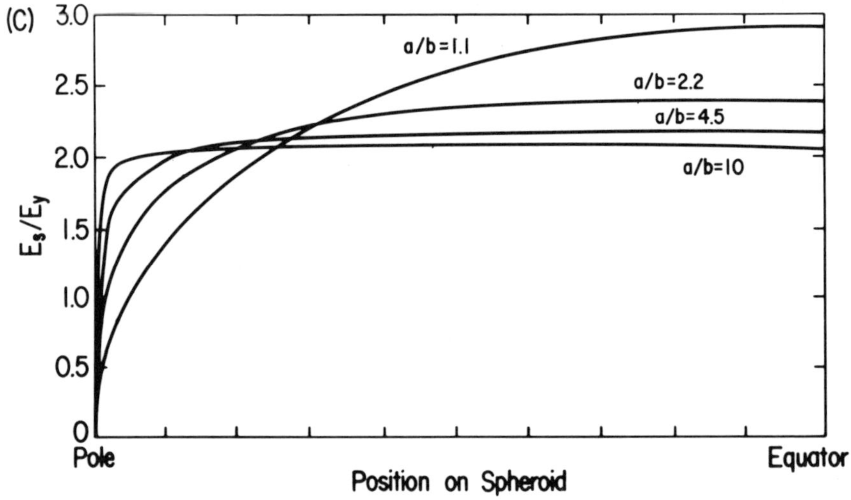

Figure 4-3c. E_s along the meridian in the xy plane is shown relative to a field applied parallel to the y-axis. Along the meridian in the xy plane E_s at the equator is intensified by a factor of 3 for a sphere (a/b = 1) and less for other shapes. Along other meridians (not shown) the fields are less, falling to zero for a meridian in the xz plane.

Because the distortions caused by man (standing upright) and four-legged animals are different, ambiguities arise in the interpretation of animal experiments when one ventures to presume the mechanism by which the field affects the organism. If the field influence is localized, then details of the surface fields may be important. On the other hand, if the effects are not local and are related to the average current density -- which is in simple correspondence to the undisturbed field -- then the field distortion may be ignored.

In experimental reports the field into which the test system is placed is usually described by the undisturbed field. Regardless of this field value, experiments on different macroscopic systems are not electrically equivalent unless the surface fields and internal currents are equivalent. It is difficult to say how serious this problem is without more insight into possible mechanisms and more data. Experiments on microscopic bodies present another problem because of the medium in which they are suspended. It is best that further discussion of this important point take place in reference to particular experimental situations.

4.7 FIELDS INSIDE THE BODY

The internal electric field is another aspect of the interaction between external fields and a conducting body. Using Ohm's Law, the internal electric field is determined by the conductivity and the current density. If only the total current is known, then we must assume a cross-sectional area in order to estimate the current density. In the example of an external field of 10 kV/m, an internal resistivity of 5 ohm-meter and an internal current of 200 μA, if the body's cross-section is taken as 0.125 m^2 (as in the study by Barnes et al., 1967), the current density is 1600 $\mu A/m^2$, or 1.6 mA/m^2. Then from Ohm's Law, the internal electric field which drives this current density is just 8.0×10^{-3} V/m, a value much smaller than the external field of 10 kV/m.

4.8 POWER LOSS INSIDE THE BODY

Energy is deposited in the tissues by ohmic heating (which is predominant over dielectric losses) (See Section 4.4). In terms of the internal electric field, E_{int}, and the body's conductivity, σ, the losses may be expressed as,

$$P = \sigma E^2_{int}$$

$$\text{or} \qquad = J^2/\sigma$$

where J is the current density in the tissue. Calculation of the heating loss using values of J and σ from Section 4.7 gives the energy absorbed per second per unit volume in the body. That is, in a 10 kV/m field, the model man absorbs about 1.3×10^{-5} watts/m^3 compared with an energy density in the undisturbed field of 4.42×10^{-8} joules/m^3. In a rat, for example, where the current density in a given external field is about 7-fold lower, the power density is about 1/50th that in man. In any case, rat or man, the heating effect is totally negligible when compared with the average metabolic rate, which in man is more than 10^3 watts/m^3.*

---

* The standard man has a resting metabolic rate of 17 cal min^{-1} kg^{-1}, equivalent to 83 watts for a 70 kg man. From the fact that the body's specific gravity is 1.07, the 70 kg body occupies a volume of 0.075 m^3 and thus the metabolic rate can be expressed as 83 watts per 0.075 m^3 or 1.1×10^3 watts/m^3. (Data source: U.S. HEW, 1970).

4.9 SUMMARY OF CONDUCTING SPHEROID MODEL

Thus, on the level of modeling the human body as a simple conducting spheroid, we find that in response to large AC electric fields, the currents which flow through the body are in proportion to the frequency, flow parallel to the external field, and change in magnitude depending upon the distance of the body to ground. We have also shown that a current flows even if there is no reference to ground, that the internal electric field strength is very much smaller than the external electric field strength, and that heating effects are negligible.

To improve this model, one should take into account the effects of the skin, which is much more highly insulating than the interior of the body, and the effect of fat and bone, which are of lower conductivity than the rest of the body. Placing the conducting body in a thin dielectric shell (skin) whose dielectric constant is considerably higher than free space, does little to change the problem. The presence of poorly conducting materials within the body does have the effect of increasing the current density over the average current densities with which we have worked. However, the preceise description of where the current flows (macroscopically) is a difficult problem (Guha _et al._, 1973; Livesay and Chen, 1974).

4.10 CURRENT FLOW INTO CELLS

How does the current flow near a cell bathed in an electrolytic solution? Because the cell membrane has such a high dielectric constant and low conductivity, most of the current (see below) flows around the cell and does not enter through the cell membrane. The precise way in which a small current does enter the cell membrane has highly interesting implications about membrane structure and can be studied from the dispersion of the dielectric constant as the frequency changes (Schwan, 1957; Hasted, 1973).

It is possible to calculate the actual current density in the interior of cells within tissues and to calculate the external electric field necessary to cause dangerous internal current densities in tissues and the external potential necessary to cause dangerous internal currents. This has been done (Schwan, 1972) for a model tissue in which spherical cells are packed to a density of about 0.8 of the volume, and suspended in a saline electrolyte. The result is that the current density inside a homogeneous spherical cell is about 0.02 times the total current density through the tissue, to within 1 order of magnitude, depending on cell size and conductivity. Consequently, he calculates the bulk tissue current density necessary to change the membrane potential V_m by an amount ΔV_m,

$$\Delta V_m = J_m/G_m,$$

where J_m is the current density across the membrane and G_m is the conductance of the membrane, which varies over a range of 10 mho/m^2 to 10^4 mho/m^2. Thus a current density of 10^{-1} A/m^2 flowing through a membrane of conductance 10^2 mho/m^2 produces a transmembrane potential of 1 mV, which is of the order of 1/50th of the resting membrane potential for many cells. By this rough estimation, we find that a transmembrane current density of 10^{-1} A/m^2 is necessary to actually influence membrane potential. As mentioned above, the bulk tissue current density which corresponds to the transmembrane current density is about 50 times larger, or 5 A/m^2.

We are now in a position to ask what electric fields are necessary to produce such a current density and the answer for a spherical model of the body is (Schwan, 1972),

$$E_{ext} = J_b/(3\omega\varepsilon_0).$$

Here, J_b is the internal body current density, E_{ext} is the external field, ω is the frequency, and ε_0 is dielectric permittivity of free space.

Taking J_b = 5A/m^2, as indicated above for a change of 1 mV in the membrane potential, we find it takes an external electric field of 5 x 10^8 V/m to cause this small change in the membrane potential.

These "order of magnitude" calculations show that exceptional electric field strengths -- greater than the breakdown strength of air allows -- are necessary to induce substantial currents into a cell. From these considerations, Schwan concludes: "It is virtually impossible to evoke dangerous current densities in the human body by external low frequency fields" (Schwan, 1972).

This conclusion may be contrasted with the following statement made by Adey, who has seen field effects in brain tissue at a strength of 7 V/m, corresponding to a tissue field strength of about 2 μV/m: "That such a weak field should be an effective stimulus is baffling in terms of classical synaptical excitation since depolarization of the post synaptic membrane in generation of an action potential transiently abolishes a gradient of about 10^5 V/m, some 10 orders of magnitude larger than these weak fields...Clearly...the unlikely possibility of...direct action on synaptic transmitter mechanisms...invites fresh consideration of more subtle membrane sensitivies" (Adey, 1974).

4.11 CELL SIZE, SHAPE, AND THE MEMBRANE POTENTIAL IN AN ELECTRIC FIELD

A more detailed consideration of the transmembrane potentials generated by external electrical fields has been given (Bernhardt and Pauly, 1973), using an ellipsoidal model for the cell. They find that the potential is dependent on the shape of the cell and at low frequencies it is linearly proportional to cell length. Their results for a long cell (100 times as long as thick) show that a potential of 100 mV is generated if the cell is placed in the field of 100 V/m. While it is not reasonable to extrapolate to a very long cell, say a 0.1 meter muscle or nerve cell, for which their formula predicts a field of 1 V/m would generate a membrane potential of 100 mV, it is still instructive to see that the membrane potential is sensitive to cell proportions and dimensions. Bernhardt and Pauly comment on this in saying that they doubt that "the high sensitivity of myelinated nerve fibers with respect to the generation of action potentials in a range of 50 - 100 Hz can be understood on the basis of our model...Probably there are other physiological mechanisms, like a threshold amount of charge, which has to cross a membrane in one-half cycle..."

4.12 SUMMARY OF ELECTRICAL DESCRIPTION OF A BODY IN AN AC ELECTRIC FIELD

In summary, we have seen that when a body is placed in an external AC electric field, currents flow through the body and to the extent that the body is uniform, the current density is uniform. The interior electric fields, necessary only to "drive" the interior current, are very much smaller than the external field impinging on the surface of the body. Currents flow through the body whether it is in close proximity to the ground or suspended in the field without reference to ground. The original field is distorted and the fields at the body surface may be intensified several fold depending on the shape, position, and orientation of the particular surface. Reasonable estimates of the properties of cell membranes and the division of current between the cell and the surrounding solution showed that only a very small fraction of the current passes through the cell membrane so that the cell membrane potential is virtually unchanged by any reasonable external currents or fields. Thus, if there is an effect directly on the cell membrane it must occur by means more subtle than a direct effect on the cell membrane potential.

4.13 ELECTROSTATIC INDUCTION

When objects in an electric field assume different potentials, the possibility exists that a short duration, high current, electric spark discharge may occur between them. For example, if a man touches a vehicle or installation which assumes a potential different from his, he can receive a substantial shock. Similarly, walking through tall grass in a strong electric field, one might sense a tingling sensation or even sharp pricks from discharges between the grass and the man's legs. These shocks are called microshocks, transient shocks, or secondary shocks, to distinguish them from primary shock hazards of conductors connected to a current source (battery or generator). They are similar to the shocks received from a doorknob after walking across a plush carpet (see Figure 2-3).

The secondary shocks are quantified by measuring the total energy in the shock discharge, usually in millijoules (mJ). The energy stored by a capacitor or capacitance C charged to voltage V is,

$$\text{Energy} = 1/2\ CV^2.$$

Representing a typical situation for a man, the discharge of a 100 pF capacitor charged to 10 kV involves 5 mJ of energy, which is not a harmful amount (q.v. Section 4.15).

The study of electrostatic induction effects in the case of an umbrella held in a strong electric field indicates that numerous discharges occur as the umbrella and hand come closer in the short moment before actual contact (Takagi and Muto, 1971). The perception of shock is due to this intermittent current and not the steady current that flows when one finally holds the umbrella firmly.

4.14 HUMAN CAPACITANCE

The appropriate human capacitance is taken to be 100 pF (Deno and Zaffanella, 1975; Korobkova *et al*., 1972), although one measurement (Barthold *et al*., 1971a) gives the capacitance for a man standing on an insulated platform 0.33 meters (13 in) above ground as about 275 pF. (Also see Section 4.5.)

4.15 DISCHARGE EFFECTS

Because the energy of the capacitive discharge is released so quickly, there is no chance of producing ventricular fibrillation which would be harmful to life. A value of 50 J for short discharges has been suggested as the danger threshold for humans (Barthold *et al*., 1971a).

The exposure to high electric field potentials also causes hair nerve stimulation due to hair erection and tingling of the body between body surface and clothing, especially in the arms (Zaffanella, 1972). This is also an electrostatic effect due to the charging of body hairs and the potential differences among body parts, clothing, etc. Zaffanella reports that these sensations become annoying to a small percentage of persons starting at field strengths of 10 kV/m. Deno and Zaffanella (1975) report that discharges begin to be objectionable at between 0.5 and 1.5 mJ, for potentials of 700 - 1200 V, assuming a capacitance of 100 pF.

Krivova *et al*. (1975b) have conducted experiments on the spark discharge pain threshold using 30 human subjects in a laboratory electric field. Each subject experienced discharges from a fingertip to a grounded object which were increased in severity until his pain threshold was reached. The data indicate that at 5.2 kV/m, 80% of the subjects were below their pain thresholds. (At present, this research is known only through the translator's abstract.)

Human sensitivity to capacitative discharges has also been measured by causing discharges from external capacitors or from the body capacitance at varying voltages (Hyltén-Cavallius et al., 1975). They note that the charge delivered is a more relevant parameter than the energy of the discharge when the discharge is over a short (microsecond) interval. Their subjects perceived discharges as small as 0.5 microcoulomb (μC) while the authors conclude that a 2.5 μC discharge to the fingertips is generally "acceptable" to their subjects.

4.16 ELECTROPHORESIS AND DIELECTROPHORESIS

When a charged body is placed in an electric field, it experiences a force which, according to Coulomb's Law, is directly proportional to the charge and the field strength. Cells or other charged, mobile entities which are in solution and subject to an electric field will move in response to the field. Electrophoresis, as this phenomenon is known, is used in biological research as a means to probe the cell membrane surface charge and influences on it due to biochemical agents or a morphological change (e.g., Mayhew, 1974).

When an uncharged object is placed in a uniform electric field, it experiences no coulombic force, but in a non-uniform field it experiences the dielectrophoretic force. Thus, neutral cells will move under the influence of a non-uniform field (AC or DC). Unlike electrophoresis, in which an alternating field causes no net motion, in dielectrophoresis, an AC field acting on a neutral body produces net movement. The dielectrophoretic force results from the polarization of the object in the field (proportional to the electric field strength, E) and the subsequent interaction of the induced dipole moment with the field (again proportional to E). Thus, the dipole moves toward the region of stronger electric field in response to the unbalanced forces exerted on it by the non-uniform field, the force is proportional to the square of the field and it depends on the polarization and conduction in both the object and the medium. A complex conductivity and complex dielectric "constant" express these properties and also represent the phase relations between the electric displacement (D), current density (J), and the imposed electric field (E) (Pohl and Crane, 1972).

In one application, dielectrophoresis allowed a separation of living and dead yeast cells (Pohl and Hawk, 1966), apparently due to changes in the cell membrane upon the death of the yeast cells. Because the dielectrophoretic force depends upon the dielectric properties of a cell, it is useful as a probe of those properties. Dielectrophoresis experiments are carried out at frequencies in the range of 10^2 - 10^9 Hz, and at relatively high electric field strengths (of the order of 10^6 V/m) (Pohl, 1973).

4.17 ELECTRIC INDUCTION BY AC MAGNETIC FIELDS

According to Faraday's Law, the presence of an alternating magnetic field induces an electrical field in a closed circuit If we let R be the radius of a circular circuit representing a loop of body tissue, then the electric field, E, induced by a magnetic field, B, in this circuit at a frequency ω is,

$$E = \frac{\omega BR}{2}.$$

Thus, at 60 Hz, a field of 10^{-4} tesla (1 G) will induce an electric field of about 2 mV/m in a circuit characterized by a radius of 0.1 meter. Corresponding to this internal electric field, in a body whose conductivity is taken as 0.2 mho/m, we have a current density of 0.4 mA/m^2. This current density is equivalent to the current density induced in a person near ground by an external electric field of 2.5 kV/m.

Thus, moderately strong magnetic fields can cause circulating currents in the body comparable in magnitude to those induced by strong electric fields. It should be noted that the direction of current flow due to an external magnetic field is perpendicular to the magnetic field, while the current induced by the electric field is parallel to the electric field. It should also be noted that the magnitude of currents induced by a magnetic field depends upon the size of the current loop and thus, for a circuitous path encompassing the dimensions of the body, the current density is greatest, whereas over very small volumes in the body, the induced current density is very small.

4.18 ELECTROMAGNETOPHORESIS

When a conducting fluid medium flows in a region in which there are orthogonal electric ($\vec{E}$) and magnetic ($\vec{B}$) fields, the Lorentz force exerted on each volume of the medium is,

$$\vec{f} = \vec{J} \times \vec{B},$$

where $\vec{J}$, the electric current density, is related to the electric field by the conductivity σ,

$$\vec{J} = \sigma\vec{E}.$$

(See Figure 4-4.)

Consequently, there is a hydrostatic pressure gradient in the medium along the direction of $\vec{f}$.

If a particle with a conductivity greater than the conductivity of the medium is suspended in the medium, the force on it is greater and it accelerates along the direction $\vec{f}$. Conversely, if the particle's conductivity is less than the medium's conductivity, the acceleration is in the opposite direction -- a case

Figure 4-4. Motion in crossed electric and magnetic fields: electromagnetophoresis

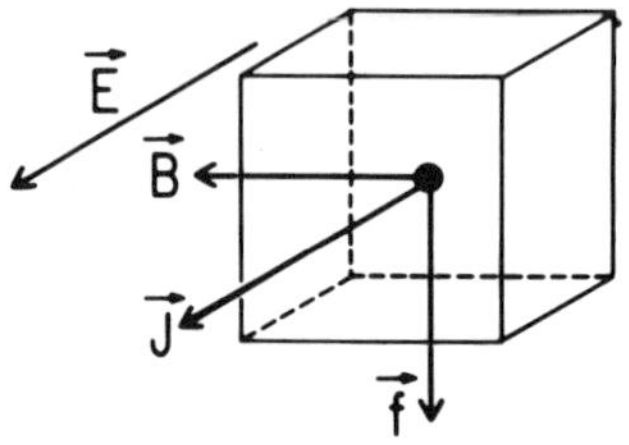

Figure 4-4a. The force per unit volume, $\vec{f}$, is due to the interaction of the crossed electric and magnetic fields on the conducting material. If unrestrained, the material will move under the action of these fields which may be AC or DC.

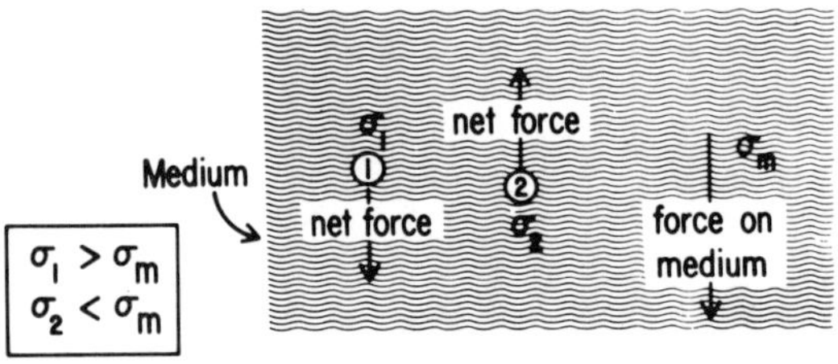

Figure 4-4b. Particle 1, whose conductivity, σ_1, is greater than the medium's has a net downward motion, and particle 2, whose conductivity, σ_2, is less than the medium's, has a net upward motion.

which is analogous to the buoyant force on an object in a liquid affected by the gravitational field (Kolin, 1960).

Because of the different electromagnetic forces on particles of different conductivity, shape, and size, it is possible to separate particles in these crossed fields, a process labeled electromagetophoresis.

The detailed analysis of the forces and velocities with which the particles move in the medium (Kolin, 1960) takes into account distortion of the current by the suspended particle, and viscous forces which arise as the particle moves through the medium. Formulae are given for a small (less than 10^{-2} cm diameter) sphere and an infinitely long cylinder oriented along either $\vec{B}$, $\vec{J}$, or $\vec{f}$. Spheroids are discussed, but the exact formulae are not available; in general, there is strong dependence on shape and orientation.

A principal advantage of electromagnetophoresis is the possibility of using AC currents and thus bypassing the troublesome electrochemical effects which occur at electrodes in DC electrophoresis.

An electromagnetophoretic device constructed by Kolin uses a magnetic field of about 6,000 G and a current density of 0.1 A/cm^2 (10^3 A/m^2) at 60 Hz.

4.19 PROBLEMS IN DOSIMETRY

Exposure to low frequency electric fields can be measured by: a) the unperturbed electric field strength existing in the region before the test body is introduced; b) the total electric current which flows in the test body; and c) the current density which flows in the body tissues. The choice of the appropriate measure and the relevance of animal experiments to humans depends in part on the mechanism of electric field influence.

As we have already seen, the electric field is distorted by the presence of the conducting body. This distortion depends on the shape and size of the body, so that humans standing erect or sitting in a chair are not comparable to animals which move on four legs.

Despite this, the unperturbed (perhaps uniform) electric field is most often used to measure the exposure with the advantages that in principle it is easy to measure and allows the intercomparison of experiments on different systems without regard for the influence of the spatial or electrical properties of the test subject.

However, if the mechanism of electric field effect is assumed to be dependent on the surface field strength, it would be necessary to use that measure. A speculative surface effects mechanism might be the irritation of superficial nerves which sense the erection of hairs, or stimulation of the muscles of the pilo-erector system. In this case, the variability from species to species might be extreme, due to different hair patterns as well as due to different surface fields. For example, the pointed ears of some animals have extraordinarily high local fields which are probably impossible to measure meaningfully.

Even in the absence of a surface effects mechanism, it may be important to know the surface fields because the greatest displacement current flows in regions exposed to the greatest field strength. In humans, the head, by virtue of its uppermost position and curved shape, is exposed to high local fields, whereas for most animals this is not so. Merely measuring total body current ignores the influence of body area, volume and mass, the direction of current flow, and the concentration in different areas (head, chest, or limbs). However, in the case of electric shock, where all these caveats also apply, the total current is commonly used to measure the shock hazards. It is necessary to accept the fundamental crudeness of this parameter because it is difficult to make measurements of the detailed current paths.

Measurement of the current density is most appropriate if one assumes a mechanism which has its effect directly at the level of certain organs, tissues or cells within the body. By specifying the current density which affects the sensitive area in one species, one could transfer the data to another species. The effects of a given current density in a mouse could thus be used to estimate the effect in humans. However, the body's non-uniformity does not permit precise estimates of actual current densities. Because the preferred current paths are not known, it is customary to assume the interior is a homogeneous conductor, with disregard for the effect of insulating and conducting materials present on the macroscopic level. Still, the imprecision caused by non-uniformity does not prevent the use of average current density as a significant measurement.

Finally, it is conceivable that properties of the animal besides the weight and current flow, such as neural mass or hair density, may be important factors in making comparisons from species to species. Until such refinements are achieved, it is best that experiments and theoretical speculations be based on the average current density to which tissues are exposed and the external electric field (which, for a given object, is simply related to the internal current density).

Because the magnetic permeability of living beings is essentially the same as for free space, the body appears "transparent" to magnetic fields. This simplifies the measurement problem for magnetic fields because the field is not perturbed when the body is introduced and all parts of the body experience the same field. Even so, magnetic field strength is not a wholly satisfactory measure because of the possible significance of currents induced by an AC magnetic field. If it were determined that current is the fundamental measure of effects these currents, although small and difficult to measure directly, would be more appropriate than the magnetic field strength. (By assuming an internal resistivity of 5 ohm-meters, the current induced over macroscopic dimensions, say 0.1 meter, is of the order of a few microamperes in a 1 G field at 60 Hz.)

4.20 ELECTRIC SHOCK CURRENTS

The human body's sensitivity to electric shocks of differing severity is described by the three current levels which, first, allow perception, second, cause involuntary muscle contraction, or third, death. At low currents, a minimum perceptible current is defined. At higher levels, a "let-go" current defines the maximum current flow at which the subject can release his grasp of the conductor, and at still higher levels, there is a minimum lethal current. These currents have been determined to within useful limits of uncertainty by tests on humans and animals and we include the data of Tables 4-5 through 4-7 for reference.

TABLE 4-5. PERCEPTION OF ELECTRIC CURRENTS*

Nature of Contact	Current at Perception
At cuts or needle puncture —	"almost too small to measure"
Tongue —	43 μA, (DC), (average)
" —	4.0μA, (DC), (minimum)
Forearm —	50 μA
Grip contact to metal cylinder —	1.0 mA, (50-60 Hz) (average)
" " —	0.4 - 0.5 mA, (60 Hz) (minimum)
1 % perception level-- holding contacts —	0.49 mA (60 Hz)
Tapping contact to plate electrode —	0.35 mA (60 Hz) (average, men)
" "	.20 mA (60 Hz) (minimum) (men and women)
1 % perception level-- touching and tapping contact —	0.13 mA (60 Hz)
Spark discharge from lineman to tower —	0.10 - 0.15 mA (60 Hz)

*Keesey and Letcher, 1969; 1970.

TABLE 4-6. LET-GO THRESHOLDS FOR ELECTRIC SHOCK*

Portion of the Population Which Can't Let Go		Threshold
50 %	(men)	15.5 mA (60 Hz)
0.5 %	(men)	9 mA "
50 %	(women)	10.5 mA "
0.5 %	(women)	6.0 mA "
(estimate)	(children)	4.5 mA "

*Keesey and Letcher, 1969; 1970

TABLE 4 7. THRESHOLDS FOR LETHAL ELECTRIC SHOCK*

ventricular fibrillation —	30 mA (60 Hz) min. for children (2 yrs.	
	60-120 mA (60 Hz) min. for adults	
thoracic tetanization —	18-22 mA (60 Hz) adults	
min. lethal threshold for —	men	9. mA
	women	6. mA
	children	4.5 mA

*Keesey and Letcher, 1969; 1970

4.21 REFERENCES

Adey, W.R. 1974. "The Influences of Impressed Electrical Fields at EEG Frequencies on Brain and Behavior." IN: "Behavior and Brain Electrical Activity," Altshuler and Burch, eds. Plenum Press, New York.

Asanova, T.P. and Rakov, A.I. 1966. "The State of Health of Persons Working in Electric Field of Outdoor 400 and 500 kV Switchyards." Hygiene of Labor and Professional Diseases 5. Translation by G. Knickerbocker available in Special Publ. #10, IEEE Power Engineering Society, Piscataway, N.J. (1975).

Barnes, H.C., McElroy, A.J. and Charkow, J.H. 1967. National analysis of electric fields in live line working. IEEE Trans. Pow. App. & Sys. PAS-86, No. 4.

Barthold, L.O. et al. 1971a. "Electrostatic Effects of Overhead Transmission Lines, Part I - Hazards and Effects." IEEE Paper 71 TP 644-PWR. Transmission and Distribution Committee of the IEEE Power Engineering Society.

Bernhardt, J. and Pauly, H. 1973. On the generation of potential differences across the membranes of ellipsoidal cells in an alternating electrical field. Biophysik 10: 89.

Cole, K.S. and Cole, R.H. 1941. Dispersion and absorption in dielectrics, I. Alternating current characteristics. J. Chem. Phys. 9: 341.

Deno, D.W. and Zaffanella, L.E. 1975. "Electrostatic Effects of Overhead Transmission Lines and Stations." IN: "Transmission Line Reference Book 345 kV and Above." Electric Power Research Institute, Palo Alto, Calif.

Guha, S.K., Khan, M.R. and Tandon, S.N. 1973. Electrical field distribution in the human body. Phys. Med. Biol. 18: 712.

Hasted, J.B. 1973. "Aqueous Dielectrics." Chapman and Hall, London. (Distributed in U.S.A. by John Wiley & Sons, New York, 1973) Chap. 8.

Heller, J.H., Wilkins, D.J., and Freeborn, J.F. 1963. Effect of radio-frequency fields on the zeta-potential of a colloidal suspension. Nature 197: 997.

Hyltén-Cavallius, N. 1975. "Certain Ecological Effects of High-Voltage Power Lines." Rept. No. IREQ-1160. Institut de Recherche de l'Hydro-Québec, Québec, Canada.

Keesey, J.C. and Letcher, F.S. 1969. "Minimum Thresholds for Physiological Responses to Flow of Alternating Electric Current through the Human Body at Power Transmission Frequencies."

Naval Medical Research Inst., National Medical Research Center, Bethesda, Md. Research Rept. No. 1, Project MR005.0801-0030B.

Kirkwood, J.G. and Shumaker, J.R. 1952a. The influence of dipole moment fluctuations on the dielectric increment of proteins in solution. Proc. Nat. Acad. Sci. 38: 855.

Kirkwood, J.G. and Shumaker, J.R. 1952b. Forces between protein molecules in solution arising from fluctuations on the dielectric increment of proteins in solution. Proc. Nat. Acad. Sci. 38: 863.

Kolin, A. 1960. "Electrophoresis: Electromagnetophoresis." IN: "Medical Physics," O. Glaser, ed., Vol. 3. Year Book Medical Publishers, Inc., Chicago, Ill., p. 261.

Korobkova, V.P. et al. 1972. "Influence of the Electric Field in 500 and 750 kV Switchyards on Maintenance Staff and Means for its Protection." CIGRE 23:06. Internat'l Conference on Large High Tension Electric Systems (28 Aug. - 6 Sept., 1972).

Krivova, T.I. et al. 1975b. "Standardization of the Intensity of an Electric Field of Commercial Frequency by the Threshold of Painful Stimulation of Electrical Discharges." Abstract by G. Knickerbocker available in Special Publ. #10, IEEE Power Engineering Society, Piscataway, N.J. (1975).

Livesay, D.E. and Chen, K-M. 1974. Electromagnetic fields induced inside arbitrarily shaped biological bodies. IEEE Trans. on Microwave Th. and Tech. MTT-22: 1273.

Mayhew, E. 1974. Interaction of polynucleotides with cultured mammalian cells, III. Initial binding at cell surfaces. Exptl. Cell Res. 86: 87.

Pohl, H.A. 1973. Biophysical aspects of dielectrophoresis. J. Biol. Phys. 1: 1.

Pohl, H.A. and Crane, J.S. 1972. Dielectrophoretic force. J. Theor. Biol. 37: 1.

Pohl, H.A. and Hawk, I. 1966. Separation of living and dead cells by dielectrophoresis. Science 152: 647.

Schneider, K.H. et al. 1974. "Displacement Currents to the Human Body Caused by the Dielectric Field under Overhead Lines." CIGRE Rept. #36-04, Internat'l Conference on Large High Voltage Electric Systems, Paris (1974).

Schwan, H.P. 1957. Electrical properties of tissue and cell suspensions. Adv. Biol. Med. Phys. 5: 147.

Schwan, H.P. 1963. Electrical characteristics of tissues. Biophysik 1: 198.

Schwan, H.P. 1972. "Biological Hazards from Exposure to ELF Electrical Fields and Potentials." Naval Weapons Laboratory Rept. NWL TW 2713.†

Schwan, H.P. and Kay, C.F. 1957a. The conductivity of living tissues. Ann. N.Y. Acad. Sci. 65, Art. 6: 1007.

Schwan, H.P. and Kay, C.F. 1957b. Capacitive properties of body tissues. Circ. Res. 5: 439.

Schwan, H.P. et al. 1962. On the low-frequency dielectric dispersion of colloidal particles in electrolyte solution. J. Phys. Chem. 66: 2626.

Schwartz, M. 1973. The theory of impedence in biological systems. J. Biol. Phys. 1: 123.

Schwarz, G. 1962. A theory of the low frequency dielectric dispersion of colloidal particles in electrolyte solution. J. Phys. Chem. 66: 2636.

Schwarz, G., Saito, M. and Schwan, H.P. 1965. On the orientation of nonspherical particles in an alternating electrical field. J. Chem. Phys. 43: 3562.

Stratton, J.A. 1941. "Electromagnetic Theory." McGraw-Hill Book Co., Inc., New York.

Takagi, T. and Muto, T. 1971. "Influences upon Human Bodies and Animals of Electrostatic Induction Caused by 500 kV Transmission Lines." Tokyo Electric Power Co., Inc., Tokyo, Japan.

U.S. Dept. of Health, Education and Welfare. 1970. "Radiological Health Handbook," rev. edition. Public Health Service Publ. No. 2016. U.S. Gov't Print. Off., Washington, D.C.

Zaffanella, L.E. 1972. Discussion of "Electrostatic Effects of Overhead Transmission Lines." IEEE Pow. App. and Sys. 91: 431-432.

CHAPTER 5

HEMATOLOGICAL AND BIOCHEMICAL EFFECTS OF ELF FIELDS

Several investigators report hematological or biochemical changes in man and lower animals exposed to electric or magnetic fields. In contradiction, other investigators find no effects. Unfortunately, the contradictions are not clearly drawn because of the widely differing experimental conditions. The studies discussed below involve both static (DC) fields and alternating (AC) fields at several frequencies within the ELF range, and they extend over a wide range of field intensities. Overall, the paucity of reliable data and the confusing picture presented by the existing data preclude any firm conclusions but, in some cases, the data do support a coherent interpretation.

5.1 THE BIOCHEMICAL TIME SCALE

This brief discussion of the time constants involved in a few biochemical and physiological processes allows some insight into the systems which might be influenced by an ELF field. The time scale on which biochemical reactions take place is uncertain and embraces a broad range. Thus, a 60 Hz variation may be slow in comparison with some events and rapid compared with others. This abstract view of chemical reactions makes it possible to reason that there may be instances where AC and DC effects are similar because, on the relevant time scale, the alternation of polarity is slow -- i.e., a "quasi-static regime." Second, and regardless of frequency, the effects may not reverse with the change in the field's direction, resulting in a non-zero net effect.

The time estimated for the incorporation of an amino acid during protein synthesis is of the order of 10 milliseconds and certainly is less than 60 milliseconds. Thus, the assembly of an entire protein may occur in about 15 seconds (Maaløe _et al_. 1966). These numbers are uncertain, but show that protein synthesis within the cell takes place on a time scale comparable to or longer than the 17 millisecond period of oscillation of a 60 Hz field.

Other characteristic times which may be relevant are those for the enzymatic reactions of electron transport at membranes, from $\ll 10^{-6}$ second to $\sim$ 10 seconds (Chance, 1972); depolarization of the membrane of a nerve cell during the production of the action potential, 1 - 2 milliseconds (Scott, 1975); and the time required to activate muscle cells, $\sim$ 2 milliseconds (Nastuk, 1974). The central nervous system rhythms indicated by the electroenceephalograph are predominantly below 20 Hz (50 milliseconds) and often are not measured beyond about 30 Hz, but there are signals at frequencies up to 60 Hz and beyond (e.g., Gedevani et al., 1974). Certain muscles (heart, intestines, etc.) are subject to rhythmic contractions, with periods of about 1 second, and later we consider the biological rhythms which have periods of many hours or days (see Table 5-1).

5.2 ELECTRIC AND MAGNETIC FIELD EFFECTS ON BLOOD COMPOSITION AND CHEMISTRY

Asanova and Rakov, 1966

The personnel at a 400 - 500 kV substation in the U.S.S.R. were given hematological tests as part of a general physical examination (see Section 6.5). This 1966 study, reported in the West in 1972, reported a number of qualitative changes in blood cell counts and cell morphology, but few numerical data were given. The average electric field strength* in the substation is reported as 7.8 - 10 kV/m and the peak field as 14.4 kV/m. The average body current is given as 115 - 125 μA and the peak body current as 230 μA. The workers, whose complaints initiated the investigation, were categorized as "maintenance personnel," exposed daily for an average of 5 hours, or as "attending personnel," exposed for an average of 2 hours per day. No control group was studied.

The following changes are reported:

- "Moderate thrombocytopenia" (reduced platelet count)
- "Moderate neutrophilic leukocytosis" (increased neutrophil count)
- "Leukopenia with neutropenia" (reduced leukocyte and neutrophil counts)

The above were "most frequently encountered." Less common observations were:

- "Lymphocytosis and monocytosis" (increases in lymphocyte and monocyte type white cells)

---

* The text of the report reads "potential field at the height of a person" and uses the units of kV, which the translator notes is an accurate transcription. If these are interpreted as free space potential measurements at a height of 1.8 meters and it is assumed the field is uniform, we arrive at the electric field strengths (unperturbed) that are given. The potentials given for the region of energized conductors are 26 kV (max) and 14 - 18 kV (average). Using 100 pF for the man-to-ground capacitance and 0.5 meter as the effective separation of the capacitor "plates," these currents, fields, and open circuit potentials are in reasonable agreement. (See Section 4.5 for details.)

TABLE 5-1. CHARACTERISTIC TIMES FOR SOME BIOLOGICAL EVENTS

Event	Time
Amino acid incorporated into a protein at the ribosome*	about 10 ms
Protein synthesis*	about 15 s
Depolarization of a neuron†	1 - 2 ms
Nerve cell impulse travels 1 cm†	0.5 to 2 ms
Activation of a muscle cell**	2 ms
Heart rate	1 s
Dominant brain waves	50 - 200 ms

* Maaløe *et al.*, 1966

† Scott, 1975 (giant squid axon)

** Nastuk, 1974

- "Tendency toward reticulopenia, moderately lowered hemoglobin and erythrocytes" (decreases in immature and mature red blood cells)

- "Retarded sedimentation rate (from 1 - 3 mm) found in 19 examinees" (longer time for red cells to settle out of serum)

- "Marked spherocytosis" in 17 of 28 studies of the red blood cells (abnormal red cell shape)

These findings are summarized in Table 5-2.

The authors do not discuss the implications of the blood findings except to call for more extensive studies of liver function and blood protein fractions.

The production of the red, white and platelet cells is performed under hormonal regulation in the bone marrow. The normal regulation of the cell population is by regulation of the production rate, so that the decreases reported suggest a hormonally-induced reduction in activity at the bone marrow. The clinical significance of the findings -- especially in the general terms in which they are reported -- is indeterminate.

"Spherocytosis" refers to the spherical shape of red cells which are normally maintained in a flattened shape by the production of ATP by glycolosis (the major metabolism of red cells). The swelling of the cells to a spherical shape occurs by the passage of water into the cells and is a clinical finding in certain hemolytic anemias (diseases in which the red cells have shortened lifespans and the red cell membrane becomes more fragile).

Retardation of the sedimentation rate (of anti-coagulated blood) indicates changes in fibrinogen or other proteins of the blood. The normal values of the male sedimentation rate are 3 - 5 mm/hour,* so that the reported changes of "from 1 - 3 mm" may be of significant magnitude.

Because numerical data are omitted, precluding evaluation of the statistical or clinical importance of the effects, and because of the lack of a control population, it is not possible to draw conclusions from this report. Despite these flaws, this study is significant in that it has raised questions of human health effects following prolonged exposure, and further epidemiological studies are indicated.

Although there is no comparable study with which to compare this Soviet report, the following report is sometimes cited as

---

* Wintrobe method -- the authors do not specify the method or the values obtained.

TABLE 5-2. CLINICAL SYMPTOMS REPORTED IN SOVIET SUBSTATION WORKERS*

1. Exposure Data

Total Group —	(41 men, 4 women)	45
Age Distribution —	less than 30 y	10
	30 - 40 y	29
	over 40 y	6
Work Experience — (in substations)	less than 1 y	9
	1 - 3 y	20
	more than 5 y	5

Average Field Exposure —

Maintenance personnel	>5 h per day
Attending personnel	<2 h per day

Electric Potentials, fields and currents —

Near energized equip't			Near disconnected equip't		
Potential	Field†	Current	Potential	Field†	Current
maximum 26 kV	14.5 kV/m	230 μA	4 kV	2.2 kV/m	35 μA
average 14-18 kV	7.8-10 kV/m	115-125 μA	2 kV	1.1 kV/m	15 μA

(continued)

---

*Asanova and Rakov, 1966.

†See text of this review for technique of determining these values which are not given in the research report.

TABLE 5-2. CLINICAL SYMPTOMS REPORTED IN SOVIET SUBSTATION WORKERS (cont'd - 2)

2. Complaints Presented —		41
headache, sluggishness, fatigue, sleepiness		37
disrupted activity of digestive tract and cardiovascular system		4
sexual weakness	1/3 of the men between 30 & 40 y	
palpitations, pains in cardiac region, irritability and poor appetite		—
3. Medical Exam Findings —		
Neural Pathology		28
Functional disruption of the CNS		26
autonomic dysfunction	(majority)	
neurasthenic syndrome	6	
Cardiovascular disease		14
neurogenic hypertension	4	
arteriosclerosis	3	
Bradycardia		12
Tachycardia		5
Hypotension		7
Hypertension		4
Electrocardiography exam		
sinus bradycardia		14
slowed atrial and ventricular conduction		10
diffuse changes in myocardium		"some"
Digestive tract disease (chronic gastritis, colicystitis)		6
4. Hematology Findings		
Moderate Thrombocytopenia Moderate Neutrophilic leukocytosis Leukopenia with neutropenia	"most frequently encountered"	
Lymphocytosis and monocytosis	"also observed"	

TABLE 5-2. CLINICAL SYMPTOMS REPORTED IN SOVIET SUBSTATION WORKERS (cont'd -

Changes in red blood cells	"more rarely"
Reticulopenia, lowered hemoglobin and erythrocytes	"tendency toward"
Retarded sed rate	19
Marked spherocytosis	17 of the 28 cases studied

a contradictory finding, although the daily exposure periods, field intensities, and work duties in the two studies are different.

Kouwenhoven et al., 1967; Singewald et al., 1972

A 9-year study was made of 11 (later 10) linemen exposed to electric and magnetic fields near 345 kV and 138 kV transmission lines in the course of maintenance operations. Of the two maintenance techniques used, the "barehanded technique," in which the lineman is at full line potential, results in stronger electric fields at the body than does the "hot stick technique," in which the lineman uses a long, insulated tool. The electric field strengths to which the men were habitually exposed while working "barehanded" on live 345 kV conductors are estimated as "around 70 kV/m." The (perturbed) body surface fields, measured with several degrees of shielding, were in the range from 20 kV/m to over 400 kV/m. The total body currents reported were in the range of 85 - 100 μA, depending on the position of the arms and the type of shielding used. Elsewhere (Kouwenhoven *et al.*, 1966), the body currents for linemen working from a bucket, or using the hot stick technique from the tower near energized 345 kV lines, are reported as 400 μA (maximum), 130 μA (minimum). The currents in tests made near 138 kV lines were somewhat less. The linemen's average daily exposure is not given. The cumulative exposures are not known since, during the study, 8 of the subjects moved from maintenance to supervisory jobs (which still exposed them to high electric fields). The magnetic field exposures are not indicated, but presumably are of the order of 1 G.

During the 9-year period, 7 extensive medical examinations were conducted at the Johns Hopkins Hospital in Baltimore. The authors report that the complete blood counts and blood chemistry tests showed "no significant changes of any kind."

No detailed data are presented and the study is flawed by the small number of men studied and the lack of controlled exposure conditions. It remains the only investigation covering such a long period of time.

The other findings, also negative, are discussed in Section 6.8.

Hauf, 1974

A laboratory study of humans was conducted at the Research Institute for Electrophysiology in Freiburg, Germany. The subjects (apparently about 100 in number) were exposed to electric fields of either 1, 15, or 20 kV/m (measured at a height of 1.35 meters). The corresponding ground level field strengths are given as 0.585, 9.1, or 11.96 kV/m. A wire mesh electrode located 2.9 meters above ground was above the subject's head as

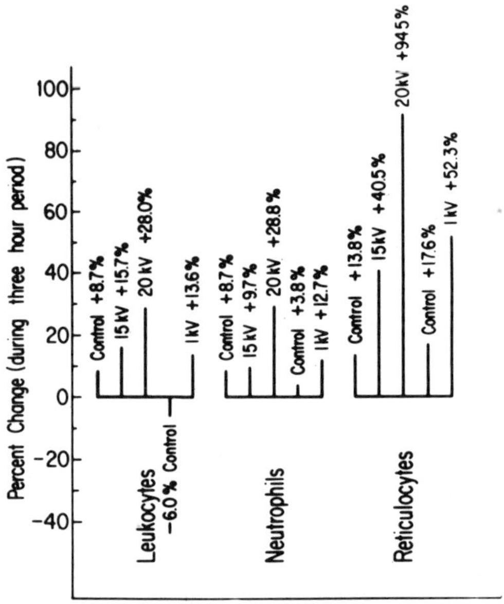

Figure 5-1. Human blood cell population changes after a 3-hour exposure to an electric field.

Source: Hauf, 1974.

he (she) sat at a table. The subject was grounded by a conductor attached to one leg.

The hematological study compared the blood counts for the exposed and control subjects as measured at the beginning and end of a 3-hour test period. These results are given in Table 5-3, and the results of physiological studies performed in other sessions are discussed in Section 6.8. The author's report does not give the number of subjects, statistical variations in the data, nor the units of measurement. In the cases where the units are apparent, they are indicated in Table 5-3.

Hauf concludes that "the leukocytes, neutrophils, and reticulocytes showed a slightly greater increase after the field exposure than in the controls. I must, however, perceive this in the sense of an irritation effect. The values remained in the normal range and a direct relation to the field strength could not be shown."*

The manner of data presentation makes it difficult to interpret the author's conclusion that the findings reflect only the "irritation" of the subject being under test for 3 hours. Apparently the exposed subjects suffered greater "irritation" than the controls -- clearly an effect if it is statistically valid. To study this question further, Figure 5-1 shows Hauf's data for the leukocytes, lymphocytes and reticulocytes (immature red cells) in the form of the percent change which occurred during the 3-hour exposure period. It is assumed that the "a" and "b" data refer to control and exposure data, respectively. The comparison of the exposed the control subjects indicates a substantial field effect.

---

* Translated by A.R.S.

TABLE 5-3. BLOOD COMPOSITION DATA FROM 3-HOUR ELECTRIC FIELD EXPOSURES *

		No Field		15 kV/m Field		20 kV/m Field		No Field		1 kV/m Field	
	(Units)	a	b	a	b	a	b	a	b	a	b
Hemoglobin	?	102.4	99.8	103.5	104.9	92.3	94.2	103.5	106.8	106.7	105.1
Erythrocytes	($x10^6/mm^3$)	5.03	4.94	5.32	5.39	4.5	4.6	5.07	5.14	5.58	5.45
Leukocytes	(no. $/mm^3$)	7140	7760	7400	8560	5463	6995	7980	7500	7800	8860
Neutrophils	(no. $/mm^3$)	4065	4417	4654	5106	3027	3399	4523	4695	4862	5478
Thrombocytes (Platelets)	(no. $/mm^3$)	216500	209420	252110	276460	329980	332870	182860	206550	264790	283060
Reticulocytes	?	5.8	6.6	4.2	5.9	5.5	10.7	6.8	8.0	4.4	6.7
Quick's Test	?	91.8	88.9	86.0	87.2	---	---	87.3	89.9	84.7	83.7

*Hauf, 1974
Units inferred from data, where reasonable.
Presumably, the "a-data" are the data before and the "b-data" after the 3-hour exposures.

If this observation is correct, it is not surprising that these particular cell populations are altered, because the body is able to regulate them on a time scale of hours, whereas the total red cell population, which did not show a field dependence, could not readily be changed in a few hours. In all cases the changes seem to be in the direction of stimulated cell production or stimulated release to the blood. Without more information on the data, it would be unwise to take the presentation of Figure 5-1 as conclusive evidence -- but it is noteworthy.

Blanchi et al., 1973a

The authors report that the white blood cell populations in mice and rats were altered significantly by exposure to a 50 Hz, 100 kV/m electric field. The exposed mice were kept in plastic cages, with metal electrodes above and below the animals, for about 55 days, during which they were exposed in a cycle of 9 hours with the field "on" and 3 hours with the field "off," for a total exposure of 1,000 hours. The control animals were housed and fed identically and precautions against corona and ozone were taken. The ozone level was monitored and was below 20 ppb. The results (Table 5-4) show significant increases in the neutrophil fraction (+98%) and eosinophil fraction (+44%) and a significant decrease in the lymphocyte fraction (-17%). The absolute counts are not reported.

Significant effects were also observed using the same 100 kV/m field for a single 6-hour exposure of rats. Once again, the eosinophil and neutrophil fractions increased and the lymphocyte fraction decreased, reaching the extreme levels 25 hours after the field was turned off. The changes in the rat white blood counts were, for the most part, reversible, returning to the control levels within 9 days.

These results may be compared with the human white cell counts already discussed. Asanova and Rakov (1966) found an increased neutrophil count among the substation workers they examined, excepting that the instances of leukopenia were accompaned by a reduced neutrophil count also. The Russian authors cited occasional lymphocytosis unlike this report, in which the lymphocyte count was lowered. In a 3-hour laboratory exposure, Hauf (1974) found that the human neutrophil counts increased to about the same degree as did the total white cell count.

Beischer et al., 1973

The effect of a low level ELF magnetic field was studied in an extensive series of tests on human subjects performed at the Naval Aerospace Medical Research Laboratory, Pensacola, Florida. A 1 G field at 45 Hz was maintained for an exposure period of about 24 hours (8 subjects plus 3 controls) or 10 hours (2 subjects, self-controlled). The subjects were confined to the test room for up to 7 days.

TABLE 5-4. EFFECTS OF A 100 kV/m FIELD (50 Hz) ON WHITE BLOOD CELLS IN RATS*

a) Differential White Cell Count in Mice Exposed to a 100 kV/m, 50 Hz Electric Field

	Neutrophils	Eosinophils	Basophils	Lymphocytes	Monocytes
Exposed	24.90±3.28 %	4.10±0.31 %	0.25±0.05 %	67.50±3.18 %	3.24±0.16 %
Controls	12.52±1.19	2.86±0.045	0.19±0.06	81.41±1.61	3.06±0.34
Difference	+12.38	+1.24	+0.06	-13.91	+0.18
% Change	+98 %	+43 %	-	-17 %	-
"p"	0.01	0.05	-	0.01	-

Mean percent of the total leukocyte count ± standard error in the mean; mice exposed to 100 kV/m for 1000^h in cycles of 9^h field on, 3^h field off.

b) Percent Change in Rat White Cells

	Neutrophils	Eosinophils	Lymphocytes
max % change	+133 %	+188 %	-25 %

Max. changes (observed 25^h after a single 6^h exposure to a 100 kV/m field) are compared to the levels 24^h before exposure

*Blanchi et al., 1973a

The daily tests of blood composition and blood chemistry were negative except for the triglyceride levels, which rose to peak values one or two days after exposure (see Figure 2-6). This result was observed in both the pilot experiment, where exposure to the magnetic field was for 9.8 hours (3 days' confinement), and in the subsequent experiment, with 22.5-hour exposure (7 days' confinement). Further evidence of an effect on the lipid metabolism was a change in the proportion of the pre-beta-lipoproteins which increased in correspondence to the triglyceride levels. The observed changes were significant at the 98% level.*

Beischer mentions an experiment on rabbits (Pautrizel *et al.*, 1972) and one on humans (de la Warr, 1967) in which the serum cholesterol levels were markedly reduced by magnetic fields (at higher field strengths in the human case). The cholesterol data are negative in the Beischer experiment. He suggests that the activity of a lipid enzyme, perhaps triglyceride lipase, is affected. The results from other parts of this investigation are discussed in Sections 6.2 and 6.3.

In a follow-up study, a 24-hour exposure to a 1 G field (45 Hz) had no effect on triglyceride levels in mouse liver (Beischer and Brehl, 1975).

deLorge, 1974

Rhesus monkeys were the subjects of this investigation into the effects of a low-level (about 9 G) homogeneous magnetic field varying at 15 or 45 Hz. The bi-weekly blood samples from the 4 monkeys (2 males and 2 females) gave essentially negative findings. Instead of the serum triglyceride increase found in man, a significant, but unrepeated, decrease was observed. This finding is complicated by the fact that the monkeys were on a partial deprivation diet and food drprivation is known to result in decreased triglyceride levels. The hematology studies, including hemoglobin, red and white cell counts, found no significant effects.

Krumpe and Tockman, 1972; Krumpe and Tockman, 1974

Human reactions to ELF magnetic and electric fields were evaluated in a survey of personnel exposed at the Project

---

* The blood chemistry analysis included complete blood count, sed rate, total protein, albumin, calcium, inorganic phosphorus, cholesterol, uric acid, creatinine, total bilirubin, alkaline phosphatase, creatinine phosphokinase, lactate dehydrogenase, glutamin oxaloaceatic transaminase, chloride, CO_2, potassium, sodium, blood urea nitrogen, glucose and serum triglyceride; also, T-3 uptake and T-4 concentration to measure thyroid function.

Sanguine Test Facility. The subjects were exposed during their jobs to a magnetic flux density (45 - 75 Hz) of 0.15 G near the antenna, which fell to 0.01 G at 10 meters from the antenna. The 45 - 75 Hz electric field was 0.06 V/m near the antenna, and 0.04 V/m at 100 meters from the antenna. A crude index of exposure was devised based upon the area of the person's job, the dates of employment, the antenna energization data, and a factor based on the job description. The longest period of exposure spanned 21 months, the shortest 7. The two medical exams, made a year apart, evaluated the health of 24 exposed personnel in the first exam, and 12 in the second. Of these, 9 could be compared with their matched controls. The hematology and blood chemistry showed no differences between control and exposed populations. An apparent decrease in the SGOT and cholesterol levels and an increase of serum triglyceride levels was an artifact of change in laboratory procedure. The data in this experiment were plentiful and subjected to extensive analysis. See Section 6.2 for other aspects of this study.

Comparative Summary of the Foregoing Reports

The foregoing data present a confused picture of the possible effect a strong 50 or 60 Hz electric field may have on blood cell counts. There are few reliable data and in addition, none of the experiments can be compared directly because of significant differences in exposure. Asanova and Rakov made a qualitative report of effects among persons exposed for many months to the strong electric fields and weak magnetic fields of a substation environment. Kouwenhoven *et al*. reported no blood changes in the ten linemen whose exposures were for shorter periods but at higher field levels. Hauf presented unclear data from a study using 50 Hz electric fields of approximately the strength found near transmission lines or substation equipment, but only for a 3-hour exposure. The blood changes reported for rats and mice exposed to a very strong 50 Hz electric field (Blanchi *et al*., 1973a) were found for both acute (6-hour) and chronic exposures (1000 hours).

Beischer's report of an elevation of serum triglyceride level among men exposed to a low level 45 Hz magnetic field appears reliable, but is as yet an isolated, preliminary finding.

Finally, the survey of Sanguine personnel (Krumpe and Tockman, 1972; 1974), showing no significant effects, involved exposure to fields much smaller than those in the other studies.

5.3 EFFECTS ON BLOOD PROTEINS AND STEROID HORMONES

Marino et al., 1974a; 1974b; 1976a; 1976b; 1976c

The biochemical effects of a continuous exposure to an ELF electric field were studied in male rats. Groups of 12 test animals (matched to 12 controls) were exposed to a 60 Hz field of 15 kV/m (nominal intensity) continuously for 30 days. The rats were exposed in small plastic cages which contained food pellets in a metal trough, a metal water tube at the top, and bedding litter. The vertical electric field was created between an electrode below the cage (energized from a transformer and sandwiched in plywood) and the cage cover, which was grounded. The nominal value for the field strength was obtained from division of the inter-electrode potential by the electrode spacing.

The 52-day-old rats were weighed and decapitated. Serum pooled from groups of 12 animals was analyzed electrophoretically for the proteins (α, β, γ and albumin fractions) and analyzed for corticosterone (the most important rat steroid) by a fluorometric technique. The experimental results from 3 runs (36 experimentals and 36 controls) show significant ($p < 0.05$) changes in the animal weights, albumin, γ-globulin and corticosterone levels (see Table 5-5a). Data from another run (12 experimentals + 12 controls) show a slight decrease in the amount of total proteins and a significant ($p < 0.05$) decrease of 10% in the rat's weights.

Possible artifacts due to animal housing, cage vibration, microcurrents and microshocks or the highly distorted field caused by the metallic feed trough (at ground potential) were considered in additional experiments using isolation pads and a plastic feed trough with a flat metal cage cover. In these and subsequent experiments, food and water consumption were monitored throughout the 30-day exposure period. Measurements of the enzyme SGOT and of the weights of the pituitary and adrenal glands were added to the experiment.

Marino et al. (1976c) report on a total of 9 experiments which showed instances of statistically significant decreases in corticosterone levels, increases in pituitary and adrenal weights, increases in levels of SGOT and albumin, decreases in water consumption, and decreases in body weights of the exposed group (see Table 5-5b,c,d). However, only the decrement in water consumption of the exposed group was regularly demonstrated at statistically significant levels ($p < 0.05$) while the other measures gave inconsistent results. The authors conclude that the overall picture indicates that the 15 kV/m electric field exposure "is productive of a physiological stress response." (Marino et al., 1976c). Some of these studies also looked for effects caused by shipment of the rats from a distant supplier and for effects dependent on the length of the acclimatization period in the laboratory, but no such effects were discovered. Another experiment examined the animals' response to a cold stress administered at the end of the exposure period. The experimental and control animals showed equal responses in this test of the competence of the pituitary-adrenal system.

Possible artifacts due to vibration and field distortion were not found. A remaining consideration is that the animals' decreased water consumption may be due to microcurrents or

a) Effect on Serum Proteins, Steroids and Weight
First Series (from Marino *et al.*, 1974a, 1974b)

		No. Rats	Wt. at 52nd day	Albumin* (%)	α* (%)	β* (%)	γ* (%)	Serum Corticosterone (μg/100 ml)
1.	Exp.	12	282.9±21.9 g	59.1±1.7	22.3±1.0	12.2±0.9	6.4±1.4	16.5±3.0
	Control	19	298.8±20.5 g	53.8±1.6	24.1±0.9	13.6±0.9	8.6±1.2	27.0±3.0
	% diff		-5% †	+10% †	-7% †	-10%	-25% †	-39%
2.	Exp.	12	274.4±15.0 g	52.4±2.8	25.7±1.0	14.6±1.1	7.5±0.8	13.5±3.0
	Control	16	282.7±16.1 g	49.2±1.0	26.9±1.0	14.7±0.6	9.2±0.7	22.6±3.0
	% diff		-3%	+6% †	-4%	-1%	-18% †	-40%
3.	Exp.	14	227.0±19.7 g	52.7±1.4	25.9±2.1	14.8±0.9	7.3±0.9	12.1±3.0
	Control	11	250.9±15.5 g	49.1±3.3	27.2±2.4	14.0±0.6	9.6±1.2	19.6±3.0
	% diff		-10% †	+7% †	-5%	+6%	-24% †	-38%
				Albumin**		Total Protein**		
4.	Exp.	12	248 ± 8.9 g	4.0 0.1 g/100ml		6.9 0.1 g/100ml		
	Control	17	276.4±15.2 g	4.0 0.1 g/100ml		7.2 0.1 g/100ml		
	% diff		-10% †	0%		-4%		

*Measurements made of pooled sera by integration under tracing of electrophoretic strip
N.B. The "% difference" here is a percent change in data which themselves are a fraction (in %). (5 strips/pool, 2 scans/strip: 10 determinations reported as "average" ± "standard deviation").

**Measurement made of pooled sera by autoanalyzer (3 determinations).

† $p < 0.05$

TABLE 5-5. 30-DAY EXPOSURE OF RATS TO 15 kV/m, 60 Hz FIELDS -- Cont'd

b) Effect on Serum Proteins, Steroids and Weight
Second Series (from Marino et al., 1976a, 1976c)

Exp. No.	Age (Days)	No. of Rats		Final Body Wt. (grams)	Serum Corticoids (μg/100 ml)	Serum Proteins (g/dl) Total	Albumin
1	(23,52)	Exper.	(14)	244.0* ± 12.6	14.6	7.4	4.4
		Control	(19)	273.1 ± 16.7	22.0	7.1	3.8
2	(24,57)	Exper.	(14)	276.5 ± 24.1	12.4	8.4	4.4
		Control	(22)	290.8 ± 27.9	18.0	6.6	3.2
3	(22,55)	Exper.	(14)	270.4 ± 14.3	10.4	8.8	4.6
		Control	(18)	277.8 ± 15.8	14.5	7.1	3.5
4	(21,51)	Exper.	(14)	228.7* ± 26.1	56.8	--	--
		Control	(11)	251.0 ± 11.3	53.4	--	--

* $P < 0.05$

N.B. Serum corticoids and albumin were measured employing the pooled sera of all rats within each group. Numbers in first column indicate age, in days, at initiation and termination of exposure. In Experiment 4, the animals were cold stressed at -13°C for 1 hour before sacrifice.

TABLE 5-5. 30-DAY EXPOSURE OF RATS TO 15 kV/m, 60 Hz FIELDS -- Cont'd

c) Effects on Water and Food Consumption and Final Body Weight (from Marino *et al.*, 1976c)

Exp. No.	Age (Days)	No. of Rats		Water Consumed (ml/rat)	Food Consumed (grams/rat)	Final Body Weight (grams)
5	(24,57)	Exper.	(15)	846* ± 68	587 ± 36	293.4 ± 17.0
		Control	(18)	940 ± 142	603 ± 40	286.7 ± 22.1
6	(22,54)	Exper.	(14)	749* ± 80	--	264.8*± 24.1
		Control	(20)	891 ± 93	--	281.0 ± 12.5
7	(25,56)	Exper.	(19)	819* ± 83	582 ± 29	289.5 ± 14.9
		Control	(21)	890 ± 104	588 ± 42	287.8 ± 18.4
8	(25,55)	Exper.	(16)	901* ± 50	542 ± 33	282.4 ± 13.3
		Control	(14)	1054 ± 84	545 ± 36	283.0 ± 12.7
9	(24,56)	Exper.	(20)	1003* ± 82	614 ± 32	286.8 ± 17.9
		Control	(20)	1099 ± 117	618 ± 43	290.2 ± 13.2
10	(24,58)	Exper.	(14)	1143 ± 157	642*± 31	294.6 ± 15.2
		Control	(16)	1202 ± 107	664 ± 17	300.4 ± 12.0

* $P < 0.05$

N.B. Numbers in first column indicate age, in days, at initiation and termination of exposure.

TABLE 5-5. 30-DAY EXPOSURE OF RATS TO 15 kV/m, 60 Hz FIELDS -- Cont'd

d) Effects on Adrenal and Pituitary Weights, Steroids, and SGOT (from Marino *et al.*, 1976c)

Exp. No.	Age (Days)	No. of Rats		Final Adrenal Weight (μg/g)	Final Pituitary Weight (μg/g)	Serum Corticoids (μg/100 ml)	SGOT (I.U.)
5	(24,57)	Exper.	(15)	180.1 ± 19.8	38.7* ± 3.2	6.8* ± 0.8	198
		Control	(18)	181.8 ± 16.0	35.2 ± 3.8	8.7 ± 1.2	194
6	(22,54)	Exper.	(14)	181.6* ± 21.1	43.9* ± 4.1	7.2 ± 1.5	188
		Control	(20)	158.9 ± 18.3	40.6 ± 3.1	7.6 ± 2.1	157
7	(25,26)	Exper.	(19)	165.3 ± 18.5	32.9 ± 3.1	--	180
		Control	(21)	168.8 ± 20.3	35.2 ± 2.6	--	157
8	(25,55)	Exper.	(16)	155.2 ± 18.4	38.0 ± 2.4	6.0 ± 0.7	137
		Control	(14)	155.8 ± 30.0	39.0 ± 2.6	6.4 ± 0.6	134
9	(24,56)	Exper.	(20)	132.1 ± 15.8	31.4* ± 2.4	9.1* ± 2.1	196
		Control	(20)	125.7 ± 14.3	29.4 ± 2.9	16.3 ± 3.8	185
10	(24,58)	Exper.	(14)	173.9 ± 20.0	31.2 ± 1.8	9.5 ± 2.0	191
		Control	(16)	179.3 ± 15.8	30.6 ± 1.8	9.7 ± 4.0	133

* $p < 0.05$

N.B. Numbers in first column indicate age, in days, at initiation and termination of exposure.

Serum corticoids were measured employing sub-pools of 2 - 3 rats within each group.

microshocks received upon touching the metallic drinking straw. The authors state that the 5 µA current which the animals receive is not a "significant factor" because water consumption was consistently the same in both groups throughout the first 14 days of exposure and showed strong differences only in the latter half of the exposure. (Food consumption, also attended by a 5 µA microcurrent, showed no differences.) The authors make this statement in the belief that if the currents were a factor, they should have an effect from the beginning. (Reference to Table 4-5 shows that in humans a 5 µA DC current can barely be detected at the tongue by some persons, although the average perception level is 43 µA. It would be interesting to know the corresponding numbers for a rat touching a source of 60 Hz current.) The question of microshocks from the drinking straw was not addressed. However, a similar argument based on the delayed onset of the effect may be made for microshocks also. No other explanation for the decrement in water consumption is given.

In the same laboratory, mice have been exposed to electric fields throughout the life cycle in a multi-generational study carried through 3 generations. Groups of 8 female and 4 male mice were exposed to either a horizontal electric field of 10 kV/m (60 Hz), a vertical field of 15 kV/m (60 Hz), or the control condition of ambient fields. The females were continuously exposed through pregnancy and statistics were kept on the number of offspring and their mortality. At 35 days of age the offspring were weighed and animals selected randomly from the first generation were mated to produce a second generation and similarly on to the third generation. The results (see Table 5-6) indicate a sharp decrease in body weights and a significant increase in mortality during the period 8 - 35 days post partum -- especially in the mice exposed in the vertical 15 kV/m field. The strongest effects were seen in the second generation, where mortality among the vertical field group was 58% at 35 days. The horizontal field group mice showed lesser effects in the first and second generations and no body weight effects in the third. In general, males and females suffered similar effects and a weighing of the second generation mice at 10 weeks disclosed that the exposed and control mice were closer in weight. The authors suggest that the differences between the horizontal and vertical field exposures may be due to position-dependent effects as well as the different field strengths. The possibility of an effect on the vertical field mice from the microcurrents incurred during eating or drinking (after weaning) is acknowledged but not resolved.

Becker is of the opinion that the electric field acts as a chronic stressor in response to which the steroid level falls, growth and development are retarded, and other pathology is more probable (Becker, 1974b). In the theory of stress advanced by Selye (Selye, 1950), and known as the "General Adaptation Syndrome," prolonged stress is a potentiator of any pre-existing or co-existing pathology and harms overall health. Becker (1974a)

TABLE 5-6. MULTI-GENERATIONAL STUDY IN MICE EXPOSED TO A STRONG 60 Hz ELECTRIC FIELD (from Marino *et al*., 1976b)

Treatment	A	B	C	D	E	F	Aver. Wt. (g) 35 d after Birth: Female	Aver. Wt. (g) 35 d after Birth: Male
1st Generation								
15 kV/m vertical field	8	5	52	10.4	4%	10%	(23) 17.0*±3.2	(22) 22.8*±4.9
10 kV/m horizontal field	8	5	47	9.4	0	2%	(21) 22.0*±1.8	(25) 23.9*±3.4
Control	8	5	61	12.2	15%	2%	(26) 24.0 ±1.5	(25) 27.0 ±2.0
2nd Generation								
15 kV/m vertical field	8	7	81	11.5	0	58%	(17) 13.6*±2.8	(15) 15.4*±2.9
10 kV/m horizontal field	8	8	92	11.5	3%	0	(43) 19.2*±2.7	(47) 22.8 ±2.8
Control	8	8	91	11.4	1%	4%	(42) 23.6 ±1.7	(44) 27.5 ±2.8
3rd Generation								
15 kV/m vertical field	8	6	54	9.0	0	35%	(19) 17.6 ±3.8	(16) 17.3*±4.7
10 kV/m horizontal field	8	8	102	12.8	14%	3%	(42) 20.9 ±2.5	(43) 23.3 ±2.9
Control	8	8	98	12.2	0	2%	(47) 19.2 ±3.3	(49) 21.7 ±3.8

Treatment	A	B	C	D	E	F	Aver. Wt. (g) 10 wks after Birth: Female	Aver. Wt. (g) 10 wks after Birth: Male
2nd Generation								
15 kV/m vertical field	8	7	81	11.5	0	58%	(16) 27.5†±2.5	(15) 29.2*±2.6
10 kV/m horizontal field	8	8	92	11.5	3%	0	(28) 27.2*±2.3	(39) 33.7*±3.0
Control	8	8	91	11.4	1%	4%	(34) 29.9 ±2.5	(32) 36.5 ±2.9

*$P < 0.001$

†$P < 0.01$

N.B. A, number of females mated; B, number of pregnancies; C, number of pups delivered; D, average litter size; E, mortality during 1st week postpartum; F, mortality during [illegible] The number of mice is given in parentheses.

says: "Chronic stress has been linked to cardiac (hypertension), renal (nephritis), gastro-intestinal (ulcers) and nervous (psychoses) diseases."

These results at 60 Hz may be compared to those in an experiment from Becker's laboratory which was performed with static fields (DC) at 2.8, 5.6, and 19.7 kV/m, using rats in the same experimental set-up (Marino *et al.*, 1974b). In these experiments, performed with both horizontally and vertically polarized fields, no effect on the growth and development (measured by weight gain) was seen, but there was a striking incidence of secondary glaucoma in rats exposed to vertical -- but not horizontal -- electric fields (see Table 5-7). This glaucoma (curiously affecting only the right eye) was explained as the exacerbation of a pre-existing uveitis commonly found in a fraction (1 - 2%) of lab rats. The dependence on polarization may be due to the randomizing effect of the rats' frequent turning in the horizontal plane, which would give cancellation for an effect which had a time constant longer than a few seconds and which was dependent on the field direction (Becker, personal communication).

The static field exposures also produced a significant change in the albumin fraction of the serum proteins and changes of varying statistical significance in the α, β, and γ globulins (see Table 5-7). The steroid levels were not assayed.

A third static field effect demonstrated that horizontal fields in the range of 8 - 16 kV/m were associated with a remarkably greater number of chromosomal abnormalities in tumor cells which were injected into live hosts (female mice) who were exposed to the fields for 14 days. No host animal cells showed chromosomal abnormalities (see Table 5-8).

The large decrements in corticosterone serum levels (plasma levels are equivalent) found in the experiment with a 60 Hz, 15 kV/m field should be viewed with an understanding of the sensitive nature of steroid assays. The corticosterone levels of rats are normally subject to large variations (up to a factor of 4) during the daily cycle. Handling, change of environment, administration of ether, traumatization, or other acute stresses can impose changes by a factor of up to 5 or more times the resting levels.

Typically, for male adult rats the trough of the daily cycle is approximately 5 μg/100 ml, occurring at 0300 hours, and the peak is about 20 μg/100 ml occurring at 1500 hours (Critchlow *et al.*, 1963).

Stress induced by handling has been observed to raise the corticosterone level from 5 μg/100 ml to a level of 20 μg/100 ml in 15 minutes (Barrett and Stockham, 1963). These data are for stress applied near the daily minimum. Higher values occur at the time of the daily maximum. Other stresses, such as moving

TABLE 5-7. EFFECTS OF 30 DAY EXPOSURE TO DC ELECTRIC FIELDS ON THE WEIGHT, GLAUCOMA INCIDENCE, AND SERUM PROTEIN LEVELS IN RATS*

a) Glaucoma Incidence and Weight Gain

Vertical Electric Field kV/m	Number of Rats	Incidence of Glaucoma	Weight Gain (Wgt at 52 d: Wgt at 22 d)
19.7	14	3	4.33
control	12	0	4.41
5.6	14	0	5.22
control	12	0	5.41
2.8	12	2	4.56
control	17	0	4.53
0.6	20	5	4.15
control	23	0	4.21

b) Serum Proteins (Albumin and the Globulins) Electrophoretic by Analysis of pooled sera. (The meaning of the figures after the " ± " is not specified.)

Vertical Electric Field (kV/m)	Albumin (%)	α (%)	β (%)	γ (%)
19.7	55.7 ± 1.7 [a]	30.1 ± 1.5 [b]	12.1 ± 0.2 [b]	2.1 ± 0
control	50.0 ± 1.9	32.9 ± 1.3	13.6 ± 0.7	3.6 ± 0
5.6	50.3 ± 1.8 [b]	28.2 ± 1.6	13.7 ± 1.0 [b]	7.8 ± 1
control	52.7 ± 1.0	28.8 ± 1.1	12.3 ± 0.7	6.3 ± 2
2.8	54.6 ± 2.7 [b]	29.1 ± 0.9	12.4 ± 1.2	3.5 ± 1
control	50.5 ± 1.1	29.1 ± 0.8	13.5 ± 1.0	6.6 ± 0

[a] $p < 0.01$ (highly significant)
[b] $p < 0.05$ (significant)

*Marino et al., 1974b. Rats were exposed continuously from 22 days of age to 52 days.

TABLE 5.8. CHROMOSOMAL ABERRATIONS IN TUMOR CELLS GROWN IN A HOST ANIMAL EXPOSED TO A DC ELECTRIC FIELD, *in Vivo*, FOR 14 DAYS*

Mature female mice were innoculated intraperitoneally with Ehrlich Ascites tumor cells. Following 14 days of exposure to horizontal DC electric fields of 8 to 16 kV/m the mouse was injected with a drug to halt cell division in metaphase thus allowing chromosome observation. Four hours later the tumor cells were removed by a needle, treated and observed microscopically.

	Experimental	Control
Number of mice employed	7	6
Total number of cells counted	350	300
Percentage of cells with abnormal chromosomes	20 ± 9.2[a]	5.1 ± 5.6
Average number of abnormal chromosomes per cell	2.0 ± 0.7[a]	0.8 ± 0.6
Percentage of cells with chromatid translocations	13.7 ± 8.8[b]	3.1 ± 2.8
Percentage of cells with isochromatid breaks	8.3 ± 5.6[c]	2.3 ± 2.7
Percentage of cells with dicentric chromosomes	1 ± 1	0

a $p < 0.005$
b $p < 0.01$
c $p < 0.05$

*Marino *et al.*, 1974b

the animal cages from one room to another, introduce changes of this order or larger (Seggie *et al*., 1974). Housing conditions (both the cage area per animal and the total number of animals) have been observed to make large changes in the basal corticosterone levels (Barrett and Stockham, 1963; del Pup and Palmes, 1971).

Another aspect of the measurement of corticosterone plasma levels is the response of a chronically stressed animal to an additional acute stress. A study of this (Daniels-Severs *et al*., 1973) disclosed that chronic crowding and cold raised the *plasma* levels at the end of the first week but had no significant effect from weeks 2 - 4. At the end of 8 weeks stress from chronic cold, but not crowding, there was a reduction in the basal levels. The response to acute ether stress was more complicated: higher response for cold stress and generally lower response for crowding-stressed rats.

The weight of rats exposed to chronic stress showed a decrement of ∿ 6% from crowding stress and ∿ 15% from cold stress (Daniels-Severs *et al*., 1973). Similar effects on body weight and steroid levels were found using a chemical agent injected into rats over a 10-week period (Bruckner *et al*., 1974).

Lastly, measurement of plasma corticosterone concentrations may be done by a number of different chemical procedures, which complicates intercomparison of data (especially data of various vintages).

Thus, the corticosterone levels in rat plasma are subject to so many influences that caution is required when interpreting data of this kind. The effects which Marino *et al*. have observed can be corroborated with similar results for other experiments (not involving fields) found in the literature. In the contrary view, it is not possible to rule out the possibility of procedural influences such as in housing of the animals, the method at sacrifice, etc., which could be reflected in the results.

Finally, the report of a weight decrement in conjunction with the steroid change makes a significantly stronger case that stress effects occurred in this experiment than if either result stood alone.

The total current through the body of a rat exposed to a 15 kV/m electric field is estimated to be about 0.7 μA with a corresponding current density of 100 $\mu A/m^2$.* The

---

* Using the tables given in Barnes *et al*. (1967) with the rat represented by an ellipsoid with semi-major axis 7.5 cm, semi-minor axis 2.5 cm and (from tables) parameter c = 7.1 cm, the current and current densities can be read from the tables in the units of μA and $\mu A/m^2$ when the tabulated figure is multiplied by 39.4 (the number of inches in a meter).

corresponding current and current density for a man in a 15 kV/m field are about 100 μA and about 800 μA/m^2. These figures for man have been compared with the actual currents measured for persons standing on ground (see Section 4.5), with the conclusion that the capacitance to ground increases the current to about 3 times the currents found in the case analyzed by Barnes *et al*.(1967). Consequently, the currents and current densities for the rats should be increased by an unknown factor to account for the increased capacitance between the animals and the electrodes.

Marino reports a value of 110 μA/m^2 for the current density (using the tables from Barnes *et al*.) and computes an internal electric field strength of about 10^{-4} V/m for the actual electric field which drives the internal current in a rat whose internal resistivity is 1 ohm-meter. The actual internal electric field strength may be greater due to the greater capacitance; by using a more appropriate resistivity, perhaps as high as 8 ohm-meter (Schwan, 1972), the internal field would be further increased by a factor of up to 8.

Udintsev and Moroz, 1974

The effects of a 200 G, 50 Hz magnetic field on steroid metabolism in the albino rat were investigated for exposures of 24 hours and 7 days. The 11-hydroxy-corticosteroid content of the adrenal gland increased by 22% after 1 day and by 33% after 1 week. Plasma levels increased by 50% after 1 day and by 55% after 7 days. Increases were found in both the proportions of the steroid bound to transcortin and in the free form, indicating an increase in the total quantity of hormone. The binding capacity of the transcortin was shown to be unchanged.

Comparison of the proportions of free steroid at 1 day (8.8 μg/100 ml) and 7 days (6.1 μg/100 ml) indicates that the exposed animals adjusted the plasma levels downward toward the control levels, an analysis that is verified by an increase in the bound fraction of the steroids over the same period of time. With the same caveats suggested above, we see evidence for a physiological effect, due here to a magnetic field of moderately great strength.

Friedman and Carey, 1972

Squirrel monkeys were exposed to a homogeneous 200 G DC magnetic field for 10 daily exposures of 4 hours each. Although no neuropathology was found upon examination of the brain, there was an indication of an elevated urinary level of the 17-hydroxy-corticoid hormones. The authors explain that the pattern of change in the steroids indicates stress effects consistent with Selye's theory of stress.

The conclusion that static magnetic fields can be stressors was previously suggested (Friedman and Carey, 1969) and an earlier study in humans (Friedman et al., 1967) showed behavioral changes indicating that a very low frequency magnetic field could also be a stressor.

* * *

There has been some research into the effects of electric or magnetic fields on cell metabolism, neurochemistry and bone growth, but little of this work was undertaken as a study applicable to ELF fields under environmental conditions. However, work in those areas, along with the other studies discussed below, contributes to a fuller picture of electrical or magnetic sensitivity in biological systems and may indicate the field or current strengths which are biologically significant.

5.4 EFFECTS ON MICROSOMAL METABOLISM

Riesen et al., 1971; Straub and Carver, 1975

Mitochondria and synaptosomes isolated from brain and liver cells were exposed to electric and magnetic fields *in vitro* as a sensitive test of the biochemical function of these organelles. The 60 Hz electric field strengths were 6.3 V/m (corresponding to a current density in the medium of 70 mA/m^2) and 155 V/m (with a current density of 18 A/m^2). Similarly, mitochondria and synaptosomes were exposed to 100 G, 60 Hz magnetic fields and the synaptosomes were also exposed at 50 G.

The rationale for studying mitochondria and synaptosomes is that an effect on membrane function might affect metabolism of adenosine triphosphate (ATP). Conceivably, an electric current effect which interfered with membrane transmission of a simple cation or more complex protein would result in a change of cell function and thus a resulting change in the production of ATP, insofar as ATP is intimately involved with all forms of cell metabolism. Mitochondrial function is measured by the amount of oxygen taken up by mitochondria suspended in a medium containing the ADP and glutamate substrate needed for the production of ATP.

The functioning of the synaptosome (or nerve ending particle) is determined by a measurement of the uptake of radioactively labeled norepinephrine.

In a preliminary experiment, Riesen demonstrated the sensitivity of mitochondria to the relatively high current levels associated with electric shock. A 60 Hz, 50 mA current (0.2 seconds in duration) destroyed mitochondrial function in rat brain cells and depressed it somewhat in the rat liver cells, depending on the shock location. The existence of a threshold was demonstrated by the observation that no effect was seen at 10 mA.

Brain mitochondria exposed to a very large electric current density ceased respiratory function after a 40 minute exposure at 18 A/m^2, although respiratory function was still viable at 15 minutes. At a current density of 70 mA/m^2 there was no effect indicating that the brain mitochondrion is current-sensitive in a region somewhere above 70 mA/m^2 but below 18 A/m^2. This experiment was conducted at 4° C, with consequently lowered chemical reaction rates.

Straub and Carver (1975) performed a similar study on rat liver mitochondria held at 0° C. No effects on the ADP and oxygen levels were seen for exposures of 15 minutes to current densities as large as 5.6 x 10^3 A/m^2.

In a related study on microsomal activity in guinea pig brain, no effects were seen on the level of membrane-bound NA-K ATPase when the microsomal preparation was incubated at 37° C for 15 minutes at current densities up to 50 A/m^2.

In Riesen's work, the exposure of brain mitochondria at 10° C to a 100 G magnetic field for 100 minutes had no effect on mitochondrial metabolism. In contrast, the synaptosomes exposed to a 50 or 100 G magnetic field for 30 minutes had a decreased norepinephrine uptake at 0° C, while at higher temperatures no such field effect occurred.

These effects on organelle function were seen at current densities very much greater than those associated with environmental exposures to electric fields. However, the demonstration that organelle function is affected by external currents or fields is, of itself, a remarkable finding. If it is assumed that the effect occurs at the mitochondrial membrane, similar results might occur at the cell membrane.

If confirmed by further experimentation, the magnetic field findings would be relevant at the levels of some environmental fields. These results are consistent with behavioral experiments in which effects of electric and magnetic fields were observed that are specific to the nervous system: the disruption of the function of the synaptosomes implies alteration of nervous function.

5.5 EFFECTS ON RESPIRATION

Greenberg and Ash, 1974; Greenberg, 1975

Respiration as a measure of the well-being of a living system was utilized in studies of 4 species of soil-dwelling animals: earthworm, slug, wood louse, and salamander. These animals lived in the environment altered by a test antenna for Project Sanguine, which was operated intermittently at 45 and 75 Hz over a 4-year period. Electric and magnetic fields in the regions where the samples were collected were measured and found to be of the order of 10 mV/m for the electric field and 0.1 - 0.3 G for the magnetic field. This experiment has been repeated several times with slight variations, and with but two exceptions the results have been negative. The wood lice were found to have a marginally significant increase in respiration in one test, but this was not confirmed in subsequent measurements. In an experiment made late in the seasonal cycle, the O_2 consumption of the red worms was, on one occasion, found to have increased very marginally.

The conclusion reached by the investigator was that "after 5 years of Sanguine operation, metabolic rate of summer or fall populations of soil animals living in the electromagnetic fields is not altered" (Greenberg, 1975).

Another experimental investigation of respiration rates showed considerable variability in rabbit tissue that was correlated with normally occurring pulses in the electromagnetic variations of the earth's field. These results (Lotmar and Ranscht-Froemsdorff, 1968; Damaschke and Becker, 1964) are cited by Persinger (1973), who goes on to note several other correlations which have been performed between geomagnetic activity and the respiration of snails. An experimental verification is reported (Lotmar _et al._, 1969) in which mouse liver tissue exposed to a VHF carrier and pulses of 30 - 100 Hz at greater than 100 mV/m showed a 42% decrease in respiration. These, and other experiments reviewed by Persinger, are from a body of literature concerning the effects of natural atmospheric changes characterized by ELF pulses of relatively low intensity. We have not made a substantial investigation of this literature, and refer the interested reader to Persinger's comprehensive review.

5.6 CALCIUM AND GABA EFFLUX IN BRAIN TISSUES

Bawin and Adey, 1975; Kaczmarek and Adey, 1974

A series of experiments which present evidence that biochemical functions related to the nervous system's operation are affected by small external electric fields was conducted at the Brain Research Institute (at the University of California, Los Angeles) by Adey and his co-workers. The most recent report (Bawin and Adey, 1975) discusses an _in vitro_ study performed on freshly-isolated chick and cat brains. The chick brains were allowed to take up calcium, labeled by radioactive Ca^{45}, for 30 minutes. They were then rinsed and exposed for 20 minutes to 5, 10 or 56 V/m AC electric fields using frequencies of 1, 6, 16, or 32 Hz. The field strengths refer to the values in the air. The efflux of Ca^{45} into a fresh solution bathing the brain was measured and, in comparison with the controls, a 15 - 20% decrease was observed at 6 and 16 Hz for the 10 and 56 V/m field strengths. The effects in the chick brains have a threshold near 5 V/m, at which field strength there were statistically insignificant "trends," and are frequency-dependent in that no significant results were seen at 1 or 32 Hz (see Table 5-9).

A similar experiment in brain tissues from the cat showed significantly decreased Ca^{45} effluxes at 6 and 16 Hz using a 56 V/m field. At this field strength the experiments at 1, 32,

TABLE 5-9. EFFLUX OF CALCIUM IONS FROM BRAIN TISSUES EXPOSED TO VARIOUS FORMS OF ELECTRIC FIELDS

A. Calcium Efflux From Brains Exposed to Electric Field in Air*

0 unchanged or insignificant
- decreased efflux
+ increased efflux

Blank: not tested
* "P" < 0.005
† "P" < 0.01

Field strength (in air)	5					10					56 V/m				
Frequency	1	6	16	32	75	1	6	16	32	75	1	6	16	32	75
Chick brain	0	0	0	0		0	-*	-*	0		0	-*	-*	0	
Cat brain	0	0	0	0	0	0	0	0	0	0	0	-*	-†	0	0

B. Calcium and GABA efflux from cat brain exposed to current from agar** electrodes.

Field strength (in tissue) : 5 V/m; pulsed, 1 ms duration, 200 pulses/second

Ca^{45} efflux: 1.29±0.06 times control efflux +

GABA efflux: 1.21±0.04 " " " +

C. Calcium Efflux in Chick brains using ELF modulated VHF carrier (147 MHZ)†

Field strength 1 - 2 mW/cm^2 at 147 MHz.

Modulation frequency (Hz)	0.5	3	6	9	11	16	20	25	35
Ca^{45} Efflux		0	+*	+*	+†	+†	+*	0	0

* Bawin and Adey, 1975

** Kaczmarek and Adey, 1974

† Bawin *et al.*, 1975

and 75 Hz were negative and at 10 V/m the cat brain tissue showed slight but not significant effects at 6 and 16 Hz (see Table 5-8).

In a previous experiment, Kaczmarek and Adey (1974) measured an increased efflux of both calcium and GABA (γ-aminobutyric acid, an inhibitory neurotransmitter) in the cerebral cortex of the cat brain. Agar electrodes were used to impose a pulsed electric field of 2 - 5 V/m, measured in the cerebral tissue. The 1 millisecond pulses were repeated 200 times per second. This is a much higher tissue-level field than in the above experiment, where the efflux was seen to decrease.

Assuming a conductivity for the cortex tissue of 0.3 mho/m, the current density is 1.5 A/m at the measured cortex potential gradient of 5 V/m. The electric field decreases with depth in the cortex. An exceptionally strong external electric field would be necessary to create the large current density required to cause this effect. These findings are relevant to research on the electrical interactions among neural cells in the brain, where the electrical complexity and sensitivity of the human body is most apparent. However, in the environmental context, the results of this particular experiment have less applicability than the foregoing experiment, because the extraordinary current density necessary in this experiment is unlikely in any external electric field exposure. Below, we discuss a related experiment in which a modulated 147 MHZ field also increased the efflux of calcium ions from chick brains studies _in vitro_.

Bawin et al., 1975

Evidence that the central nervous system reacts in a frequency-selective way to external fields is provided in this experiment, where a VHF (147 MHz) carrier of low power (1 mW/cm2) modulated at ELF frequencies (0.5, 3, 6, 9, 11, 16, 20, 25, and 35 Hz) was used to irradiate the brains of chicks _in vitro_. The brains of freshly-sacrificed chicks were incubated for 30 minutes to allow the uptake of a medium labeled with radioactive $^{45}Ca^{++}$, so that the efflux of calcium, measured by the $^{45}Ca^{++}$, could be determined. After rinsing, the brains were exposed for 20 minutes to the combined VHF-ELF fields, after which the medium was assayed for the radioactive calcium content. At each frequency, 10 brains were exposed with 10 unexposed control brains. In a comparison with brains not exposed to the VHF field, there was no significant change in Ca^{++} efflux for exposure to the unmodulated 147 MHz carrier or the carrier modulated at 0.5 or 3 Hz. However, statistically significant results ($p < 0.05$ or 0.01) were found at modulation frequencies of 6, 9, 11, and 16 Hz, while at 25 and 35 Hz the results were again not significantly different from control values. The peak effect (20% greater efflux) occurred at 16 Hz with smaller effluxes as frequency increased or decreased (see Figure 4-2 and Table 5-10).

TABLE 5-10. EFFECTS OF AMPLITUDE-MODULATED VHF FIELDS ON $^{45}Ca^{2+}$ EFFLUX FROM THE ISOLATED CHICK BRAIN*

	Relative $^{45}Ca^{2+}$ Efflux	
	Normal Brain	Poisoned Brain
Control, no field	100.0 ± 4.0(15)	96.7 ± 3.6(15)
Unmodulated field	103.7 ± 6.0(10)	102.3 ± 5.3(10)
0.5-Hz modulation	100.5 ± 4.6(10)	98.7 ± 5.0(10)
16-Hz modulation	114.2 ± 6.4(10) †	118.9 ± 7.7(10) †

TABLE 5.10. EFFECTS OF AMPLITUDE-MODULATED VHF FIELDS ON $^{45}Ca^{2+}$ EFFLUX FROM BIOLOGIC TISSUES*

	Relative $^{45}Ca^{2+}$ Efflux	
	Brain	Muscle
Control, no field	100.0 ± 5.4(10)	100.0 ± 5.8(10)
6-Hz modulation	115.0 ± 4.2(10) †	105.1 ± 9.8(10)
16-Hz modulation	119.1 ± 6.2(10) †	100.1 ± 3.9(15)

† $p < 0.05$

The data are referred to the mean control value of 100 and are given ± the standard error in the mean. The numbers in parentheses are the numbers of determinations.

*Bawin *et al.*, 1975

Portions of the experiment (at 0.5, 16 Hz and without modulation) were repeated using brains whose metabolism had been halted before the exposure by poisoning with sodium cyanide. The same results were found with these dead brain cells as for the live cells, suggesting that the increased calcium efflux is from outside the cell -- that is, from the counterion layer of the greater cell membrane, according to Adey's hypothesis.

A preliminary experiment on muscle tissues determined that there was no change in the calcium efflux due to external fields at 6 or 16 Hz modulation frequencies or with an unmodulated carrier. Thus far, the observed changes are peculiar to neural tissue.

Finally, it may be significant that the most effective frequencies in this experiment, 11 and 16 Hz, are constituents of the EEG of the aroused chick brain.

5.7 EFFECTS ON PROTEIN SYNTHESIS

We now consider a number of experiments which show that living systems respond to electrostatic fields (DC).

Harrington and Becker, 1973

Frog erythrocytes were exposed to a non-uniform static electric field that produced a current of between 250 and 1,000 pA for the purpose of measuring changes in protein synthesis within the cell. The frog erythrocyte is unlike the human cell in that it contains a nucleus. In this experiment the erythrocytes, suspended in a cell medium, were exposed to the currents in a non-uniform field between platinum electrodes. The tests continued for 8 hours in the presence of radioactively-labeled amino acids, enabling measurement of the incorporation of the amino acids into proteins manufactured by the erythrocytes during the exposure period. The results showed no DNA synthesis in either the controls or the exposed cells. However, there was evidence that RNA had been synthesized by the exposed cells while the controls, as would be expected of mature red blood cells, did not demonstrate protein synthesis. If a uniform electric field was established through the medium, the effect was not observed.

This experiment demonstrated that a cell which normally does not synthesize RNA can be stimulated to do so by an external electric current. This implies that the erythrocyte nucleus, normally inhibited from expressing its ability to make RNA, can be affected and the inhibition mechanism defeated. The results of this experiment cannot be extrapolated from DC to AC without greater understanding.

Bassett and Hermann, 1968

Further evidence that electrostatic fields can affect the synthesis of macromolecules is provided by an experiment made on fibroblast tissues placed in a field of 10^5 V/m. After a 14-day exposure, the effect on synthesis of DNA and collagen -- the protein which forms connective tissue -- was detected by a radioactive labeling technique. The continuous electrostatic field increased DNA production by 20%. A field interrupted once per second increased DNA synthesis, again by 20%, and collagen synthesis by 100%. A dependence upon the field strength was noted.

The authors relate these *in vitro* findings to the control of the proliferation of connective tissue that is apparently regulated by bone and cartilage through their piezoelectric properties.

5.8 EFFECTS ON IMMUNOLOGICAL ABILITY

Fischer, 1973

Electric fields in the range of 2400 kV/m were found to have a beneficial effect on mice as measured by their activity, rate of liver respiration, and ability to form antibodies. In contrast, mice which were deprived of any electrostatic fields by being enclosed in a Faraday cage showed opposite results. This research was designed to investigate the importance of the ambient electrostatic field (120 V/m) to human health and contains the suggestion that immunological preparedness could be increased by increasing the field as a clinical practice.

5.9 EFFECTS ON ATP METABOLISM

In the Soviet Union there has been some experimentation at frequencies that are somewhat above the ELF range. Two such experiments which demonstrate biochemical changes are discussed below on the basis of English language abstracts of the Russian work.

Kolodub and Yevtushenko, 1974; 1973

In the first experiment at 7 kHz, a pulsed electromagnetic field demonstrated that morphological changes in the rat liver were correlated with and preceded by a change in metabolism. Up to 15 exposures over a 6-month period at an unspecified field strength disrupted the liver cell metabolism so that the ATP content was decreased with a concurrent increase in the associated compounds ADP and AMP, which normally are phosphorylated to ATP in the course of cell metabolism. A decline in the synthesis of urea and other biochemical changes is also reported.

Skeletal muscle of a rat exposed to a low frequency (7kHz) pulsed EM field at a field strength of 72 and 25 kA/m also demonstrated a decrease in ATP and gave other evidence of biochemical disruptions.

5.10 EFFECTS ON BONE GROWTH AND FRACTURE REPAIR

Bone is an electrically active medium where normal growth and development as well as the repair of fractures are regulated* by an electrical potential developed within the bone. This discovery has generated research into the major structural elements of bone: the long protein molecule, collagen and the crystalline mineral, apatite.

Piezoelectric materials have the property that they generate a potential according to the stress applied to them, and conversely, applying an electric potential induces stress. The piezoelectric properties of organic materials have been studied extensively by Fukada (1974). The piezoelectric properties of bone have been related to the growth and development of bone and to its repair after fracture. In this connection, a great deal of research has been conducted, of which we shall emphasize research in which external electric fields or currents have been used.

Bone Growth in Electric Fields

McElhaney *et al.*, 1968: That bone formation is stimulated in response to an applied mechanical stress has been known. This study undertook to determine if bone formation also could be affected by the piezoelectric response of collagen to an applied electric field. In this experiment the femurs of 48 male rats were exposed to various electric fields for a period of 20 days. Fields were applied to one leg using a copper plate imbedded in a plaster cast under 4 conditions: 1) a sham field (control group); 2) a DC potential of 100 V across the plates; 3) an AC signal of 200 V at 3 Hz; or 4) a 200 V signal at 30 Hz. The authors estimate that these electric fields correspond to a stress of approximately 3,000 psi. The electric field strength is estimated as perhaps 10^4 V/m. The mechanical properties of the rat femurs showed that the 30 Hz field was most effective in producing increased bone growth and strength. Based on the data from the 24 rats which survived and also showed a normal weight gain during the 28-day experimental period, the authors' conclusion is that when bone is immobilized in a cast changes occur in its structure rather than its composition

---

* Becker finds that despite the electrical activity of bone itself, growth will not occur unless an electrical pathway for DC potentials is intact. This DC path is apparently associated with the nerves but Becker has shown that the ability to transmit an action potential along neurons is not essential to bone growth (Becker, 1974a).

and these structural changes can be lessened with an external electric field.

Bone tumors were found in 8 of the animals, 5 of which were exposed at 30 Hz, 2 at 3 Hz, and 1 to the DC field. No tumors were found in the unexposed femurs. No identification of the tumors was made. Despite the fairly extensive experimentation in electrically stimulated bone growth, this is the only report of tumor growth related to the use of electric fields. One cannot say whether this is because no other experiment applied similar strong fields, or because of error in this experiment.

Lavine *et al.*, 1972: More frequently bone growth is affected by the application of DC electric currents directly to the bone using an electrode. In this report, a human bone disease was successfully treated with an artificial electric current. A 14-year-old boy afflicted with pseudarthrosis since birth fractured his tibia and there was no union of the broken bone after a period of 30 months. Two platinum electrodes were placed in holes drilled into the bony tissue on either side of the defective portion of the bone. An external battery (3 V) in series with a resistance provided a current of 4 μA through the bone tissue. The potential difference across the electrodes was 0.55 V. After 125 days of electrical treatment, significant structural development occurred at the defect portion of the bone, indicating that the bone union was beginning. Microscopic analysis of bone tissue confirmed that collagen was forming and osteogenesis occurring in the region where formerly it had not.

The authors comment: "It is clear that electrical potentials play an important role in directing the architectural and structural development of bone."

Norton, 1974: In this report, a 5 Hz electric field was introduced across rooster chick legs using electrodes. The peak electric field values were between 8.1×10^4 and 2.7×10^4 V/m. The 30-day exposures caused a gross change in bone morphology and in the pattern of growth, although no difference was seen among the various levels of the electric field.

Bone Growth Using a Pulsed Magnetic Field

Bassett *et al.*, 1974a; 1974b: In this experiment, a pulsed electric current was passed through coils to induce an electrical voltage by action of the magnetic field in the bone tissue. The coils of wire were placed next to the surgically fractured legs of adult beagles. The currents were pulsed at a 1 or 65 Hz repetition rate and caused electric fields in the range from 10 mV/m to 5 V/m. The strength of the bone 28 days after treatment began was measured using a mechanical testing device to determine stiffness and the vicsoelastic properties. Bones healed under the influence of the 1 Hz pulses showed no greater strength than the untreated fractures, but those healed under the 65 Hz pulsed field did show significantly greater strength ($p < 0.07$). In addition to the increased repair in the experimentals, the structure in the callus of the exposed bone was more highly organized.

Micrographs and X-rays which show the osteogenetic effect of the stimulating fields are presented (Bassett *et al.*, 1974b).

Electrical Influence on the Mitotic Rate in Bone

Rinaldi *et al.*, 1974: The rate of mitotic division of bone cells stimulated by a small (4 μA) electrical current was measured. The bone current was provided by electrodes placed into the bone and the defect in the bone was provided by a hole drilled in the region between the two electrodes. The concentration of tritium labeled thymidine showed increased mitosis of the osteogenic cells. Healing in the defect area improved, again demonstrating that an electric current can stimulate bone growth.

Polarity Effect

Klapper and Stallard, 1974: This experiment was performed to investigate the relative efficacy of unipolar or bipolar pulses in stimulating bone growth. It was found that although a pulsed DC signal produced a highly significant change in bone growth compared with the controls, pulses which alternated in polarity were even more effective.

Microscopic Evidence

Lavine *et al.*, 1974: Electron micrographs of human bone tissues before and after electrical stimulation of healing are presented with an analysis which finds that after electrically stimulated promotion of bone healing, the fibroblast cells show those signs of disintegration which are associated with the formation of a strong bone union. The authors discuss how the aggressive disintegration of the fibroblast cell constituents, particularly the mitochondria, may be a part of the process of forming the bony matrix, concluding that: "Based on the electron microscope evidence presented here...the mechanics of tricalcium phosphate release into the extracellular matrix is via cell membrane contact, which initiates a dissolution of the membranes. Another and related route of releasing tricalcium phosphate may be simply the breaking down of vacuoles that have become free in the extracellular matrix."

Other experiments have demonstrated that small electrical currents in the microampere range stimulate the uptake of strontium by bone (Klems *et al.*, 1975; Hambury *et al.*, 1971).

The reports cited above are a sample from an extensive and growing literature which provides ample evidence of electrical effects on bone growth, although the biophysical principles (piezoelectric properties of collagen), biochemical processes, and the evidence of microscopic examination have not yet been unified. At a recent conference, "Electrically Mediated Growth Mechanisms in Living Systems," held in New York in 1973, many papers in this field were given and have since been published in Volume 238 of the Annals of the New York Academy of Sciences.

5.11 SUMMARY

In summary, the evidence that environmentally relevant field strengths can affect blood composition, biochemical synthesis, or metabolism comes from a very few, scattered, experiments. The experimental reports of an AC electric field effect are: 1) the change in plasma protein and steroid levels and in the weight of rats exposed continuously at 15 kV/m (60 Hz); 2) the alteration of white blood cell counts observed in mice and rats at 100 kV/m (50 Hz) and in humans at 15 and 20 kV/m; and 3) the altered calcium efflux from chick and cat brains exposed to a field (in air) of 10 V/m or 56 V/m (6 and 16 Hz). In the latter experiment, no alteration occurred at 1, 32, or 75 Hz, and it is doubtful that the effect is metabolic.

Magnetic field effects were seen in the elevation of serum triglyceride levels in men exposed for 10 or 22 hours at 1 G (45 Hz), metabolic decreases in nerve cell organelles (synaptosomes) exposed at 50 and 100 G (60 Hz), and steroid changes in monkeys exposed at 200 G (DC).

Other biochemical effects are reported in experiments involving the electric field influence on brain tissue, cell organelles, frog red blood cells, and fibroblast cells in culture but at field strengths much greater than those encountered environmentally. Cultured rat tissues exposed to a pulsed electromagnetic field and brain tissue exposed to an ELF-modulated radio wave also showed biochemical effects. Finally, there are many reports to show that DC, and perhaps AC, currents have an important role in bone growth and repair, and one report of tumors related to electrical stimulation of bone.

The hematological and biochemical effects attributed to 50 - 60 Hz electric fields are uncertain because there is only the evidence from two research groups working with mice or rats (Marino *et al.*, 1974a,c; 1976a,b,c; and Blanchi *et al.*, 1973a), one sketchy medical report (Asanova and Rakov, 1966) and one inadequately reported test on humans (Hauf, 1974). This is too little on which to base an evaluation, but certainly the serious implications of these findings requires that these effects be thoroughly examined by further experimentation.

Unfortunately, excepting the studies in bone and, to a lesser extent, the neural studies, none of the problems has been adequately investigated. Thus, the reported effects have not yet been corroborated by an independent replication, nor has the experimental work advanced to the point where response thresholds, a dose-response relation or frequency dependence are known. Lacking these developments, the results must be characterized as unproven. The calcium efflux study in neural tissue is excepted insofar as a threshold (at about 10 V/m for the chick brains) and a frequency dependence (no effects at 1 or 32 Hz) have been reported.

5.12 REFERENCES

Asanova, T.P. and Rakov, A.I. 1966. The state of health of persons working in electric field of outdoor 400 and 500 kV switchyards. Hygiene of Labor and Professional Diseases 5. Translation by G. Knickerbocker available in Special Publ. #10, IEEE Power Engineering Society, Piscataway, N.J. (1975).

Barnes, H.C., McElroy, A.J. and Charkow, J.H. 1967. Rational analysis of electric fields in live line working. IEEE Trans. Pow. Appl. & Sys. PAS-86, No. 4.

Barrett, A.M. and Stockman, M.A. 1963. The effect of housing conditions and simple experimental procedures upon the corticosterone level in the plasma of rats. J. Endocrin. 26: 97.

Bassett, C.A.L. and Hermann, I. 1968. The effect of electrostatic fields on macromolecular synthesis by fibroblasts in vitro. (Abstract) J. Cell. Biol. 39: 9a.

Bassett, C.A.L., Pawluk, R.J. and Pilla, A.A. 1974a. Augmentation of bone repair by inductively coupled electromagnetic fields. Science 184: 575.

Bassett, C.A.L., Pawluk, R.J. and Pilla, A.A. 1974b. Acceleration of fracture repair by electromagnetic fields: a surgically noninvasive method. Ann. N.Y. Acad. Sci. 238: 242.

Bawin, S.M. and Adey, W.R. 1975. "Effects of Weak Low Frequency Electric Fields on Calcium Efflux from Isolated Chick and Cat Brain." Abstract of paper presented at Annual Meeting of the U.S. Nat'l Committee of the Internat'l Union of Radio Science (URSI), Boulder, Colo.

Bawin, S.M., Kaczmarek, L.K., and Adey, W.R. 1975. Effects of modulated VHF fields on the central nervous system. Ann. N.Y. Acad. Sci. 247: 74.

Becker, R.O. 1974a. The basic biological data transmission and control system influenced by electrical forces. Ann. N.Y. Acad. Sci. 238: 236.

Becker, R.O. 1974b. Prepared testimony before State of New York Public Service Commission, Case 26559.

Beischer, D.E. and Brehl, R.J. 1975. "Search for Effects of 45 Hz Magnetic Fields on Liver Triglycerides in Mice." Naval Aerospace Medical Research Laboratory, Pensacola, Fla. Rept. NAMRL-1197.

Beischer, D.E., Grissett, J.D. and Mitchell, R.E. 1973. "Exposure of Man to Magnetic Fields Alternating at Extremely Low Frequency." Naval Aerospace Medical Research Laboratory, Pensacola, Fla. Rept. NAMRL-1180. Available from NTIS as AD 770140.†

Blanchi, D. et al. 1973a. Exposure of mammalians to strong 50 Hz electric fields. 1. Effects on the proportion of the different leukocyte types. Arch. Fisiol. 70: 30.

Bruckner, J.V., Khanna, K.L. and Cornish, H.H. 1974. Polychlorinated biphenyl-induced alteration of biologic parameters in the rat. Toxicol. Appl. Pharmacol. 2: 189.

Chance, B. 1972. "Kinetic, Thermodynamic and Computer Simulation Studies of Site III Electron Transport and Energy Coupling as Studied at Normal and Subzero Temperatures." IN: "The Molecular Basis of Electron Transport," J. Schultz and B. Cameron, eds. Academic Press, New York, pp. 65-95.

Critchlow, V. et al. 1963. Sex difference in resting pituitary-adrenal function in the rat. Am. J. Phys. 205: 807.

Damaschke, V.K. and Becker, G. 1964. Korrelation der Atmungsintensitaet von Termiten zu Aenderunger der Impulsfolgefrequenz der Atmospherics. Z. Naturforschg. 19: 157-160.

Daniels-Severs, A. et al. 1973. Effect of chronic crowding and cold on the pituitary-adrenal system: responsiveness to an acute stimulus during chronic stress. Pharm. 9: 348.

deLorge, J. 1974. "A Psychobiological Study of Rhesus Monkeys Exposed to Extremely Low Frequency-Low Intensity Magnetic Fields." Naval Aerospace Medical Research Laboratory, Pensacola, Fla. Rept. NAMRL-1203. Available from NTIS as AD 000078.†

de la Warr, G.W. and Baker, D. 1967. "Biomagnetism: Preliminary Studies of Magnetic Fields on Living Tissue and Organs in the Human Body." de la Warr Laboratories, Ltd., Oxford, England.

del Pup, J. and Palmes, E.D. 1971. Effect of housing conditions on corticosterone levels in mice. Arch. Environ. Health 22: 493.

Fischer, G. 1973. Die bioklimatologische Bedeutung des elektrostatischen Gleichfeldes (The bioclimatological importance of the constant electrostatic field). Zbl. Bakt. Hyg., I. Abt. Orig. B 157: 115. (In German with English abstract)

Friedman, H., Becker, R.O. and Bachman, C.H. 1967. Effect of magnetic fields on reaction time performance. Nature 213: 949.

Friedman, H. and Carey, R.J. 1969. The effects of magnetic fields upon rabbit brains. Physiol. Behav. 4: 539.

Friedman, H. and Carey, R.J. 1972. Biomagnetic stressor effects in primates. Physiol. Behav. 9: 171.

Fukada, E. 1974. Piezoelectric properties of organic polymers. Ann. N.Y. Acad. Sci. 238: 7-25.

Gedevani, D.M. et al. 1974. Physical and mathematical simulation of the 55 c/sec bioelectric rhythm produced by the olfactory bulb. Int. J. Bio-med. Comput. 5: 1.

Greenberg, B. 1975. "Oxygen Consumption and Respiratory Quotient in Five Animal Populations Naturally Exposed to Sanguine Electromagnetic Fields." Univ. of Illinois - Chicago Circle.†

Greenberg, B. and Ash, N. 1974. Impact of extremely low frequency electromagnetic fields on soil arthropods in nature. Env. Entymol. 3: 845-853.†

Hambury, H.J., Watson, J., Sivyer, A. and D.J.B. Ashley. 1971. Effect of microamp electrical currents in bone in vivo and its measurement using strontium-85 uptake. Nature 231: 190-192.

Harrington, D.B. and Becker, R.O. 1973. Electrical stimulation of RNA and protein synthesis in the frog erythrocyte. Exp. Cell. Res. 76: 95.

Hauf, R. 1974. Wirkung von 50-Hz-Wechselfeldern auf den Menschen (Effects of 50 Hz alternating fields on man). Sonderdruck aus Elektrotechnische Zeitung-b 26: 381.

Kaczmarek, L.K. and Adey, W.R. 1974. Weak electric gradients change ionic and transmitter fluxes in cortex. Brain Res. 66: 537.

Klapper, L. and Stallard, R.E. 1974. Mechanism of electrical stimulation of bone formation. Ann. N.Y. Acad. Sci. 238: 530-542.

Klems, H., Venohr, H., and Wiegert, M. 1975. Stimulierung des Langenwachstums von Rohrenknochen durch elektrischen Gleichstrom. Szintigraphische Untersuchungen und der Kaninchentibia. Arch. Orthop. Unfall-Chir. 81: 285.

Kolodub, F.S. and Yevtushenko, G.I. 1973. Characteristics of skeletal muscle metabolism in rats under effect of pulsed electromagnetic field of low frequency. Ukr. Biokhim Zh. 45: 356. Abstract #44287 in Bio. Abstracts 59: 4734 (1975).

Kolodub, F.S. and Yevtushenko, G.I. 1974. Metabolic disorders and the liver function under the effect of a low frequency pulsed electromagnetic field. Gig. Tr. Prof. Zabol. 2: 11. Abstract # 10998 in Bio. Abstracts 59: 1174 (1975).

Kouwenhoven, W.B. et al. 1967. Medical evaluation of man working in AC electric fields." IEEE Trans. Pow. App. & Sys. PAS-86, No. 4.

Krumpe, P.E. and Tockman, M.S. 1972. "Evaluation of the Health of Personnel Working near Project Sanguine Beta Test Facility from 1971 to 1972." Naval Medical Research Unit No. 4, Great Lakes, Ill.†

Krumpe, P. E. and Tockman, M.S. 1974. "Evaluation of the Health of Personnel Working near Project Sanguine Beta Test Facility from 1971 to 1972." IN: "Biologic and Clinical Effects of Low Frequency Magnetic and Electric Fields." J. Llaurado *et al.*, eds. Charles C. Thomas, Springfield, Ill.

Lavine, L.S. *et al.* 1972. Electric enhancement of bone healing. *Science* *175*: 1118.

Lavine, L.S. *et al.* 1974. Clinical and ultrastructural investigations of bone healing. *Ann. N.Y. Acad. Sci.* *238*: 552-563.

Lotmar, R. and Ranscht-Froemsdorff, W.R. 1968. Intensitaet der Gewebeatmung und Wetterfaktoren (Korrelation des QO_2 von Kaninchenhaut und Atmospherics). *Z. angew. Bader.-u. Klima-heilk.* *15*: 1-10.

Lotmar, R., Ranscht-Froemsdorff, W.R., and Weise, H. 1969. Daempfung der Gewebeatmung (QO_2) von Mausleber durch kuenstliche Impulsstrahlung. *Int. J. Biometeor.* *13*: 231-238.

Maaløe, O., Maaløe, N.B. and Kjeldgaard, N.O. 1966. "Control of Macromolecular Synthesis." W. A. Benjamin, Inc., New York.

Marino, A.A. *et al.* 1974a. "Effect of Continuous Exposure to Extremely Low Frequency Electric Fields on Rats." Veterans Admin. Hospital and Upstate Medical Center, Syracuse, New York.

Marino, A.A. *et al.*, 1974b. Electric field effects in selected biologic systems. *Ann. N.Y. Acad. Sci.* *238*: 436.

Marino, A.A., Berger, T.J., Austin, B.P., Becker, R.O., and Hart, F.X. 1976a. Evaluation of electrochemical information transfer system. I. Effect of electric fields on living organisms. *J. Electrochem. Soc.* *123*: 1199-1200.

Marino, A.A., Becker, R.O. and Ullrich, B. 1976b. The effect of continuous exposure to low frequency electric fields on three generations of mice: a pilot study. *Experientia* (Basel) *32*: 565-566.

Marino, A.A., Berger, T.J., Austin, B.P., Becker, R.O., and Hart, F.X. 1976c. *In vivo* bioelectrochemical changes associated with exposure to extremely low frequency electric fields. *Bioenergetics and Bioelectrochemistry* (in press).

McElhaney, J.H., Stalnaker, R. and Bullard, R. 1968. Electrical fields and bone loss of disuse. *J. Biomech.* *1*: 47.

Nastuk, W.L. 1974. "Neuromuscular Transmission." IN: "Medical Physiology," 13th ed., V. Mountcastle, ed., C. V. Mosby Co., St. Louis, Mo.

Norton, L.A. 1974. *In vivo* bone growth in a controlled electric field. *Ann. N.Y. Acad. Sci.* *238*: 466-477.

Pautrizel, R. *et al.* 1972. Influence of electromagnetic waves and magnetic fields on the lipid modifications induced in the rabbit by feeding a hypercholesteremic diet. (In French). *C.R. Acad. Sci. Paris* (*D*) *274*: 488.

Persinger, M.A. 1973. Possible cardiac driving by an external rotating magnetic field. *Int. J. Biometeor.* *17*: 263-266.

Riesen, W.H. *et al.* 1971. "A Pilot Study of the Interaction of Extremely Low Frequency Electromagnetic Fields with Brain Organelles." ITT Research Institute, *Tech. Memorandum No. 3, IITRI Project E 6185*.†

Rinaldi, R., Shamos, M. and Lavine, L. 1974. Uptake of tritiated thymidine during electrical stimulation of induced cortical bone defects. *Ann. N.Y. Acad. Sci.* *238*: 307-313.

Schwan, H.P. 1972. "Biological Hazards from Expsoure to ELF Electrical Fields and Potentials." Naval Weapons Laboratory *Rept. NWL TW-2713*.†

Scott, A. 1975. Electrophysics of a nerve fiber. *Rev. Mod. Phys.* *47*: 487.

Seggie, J. *et al.* 1974. Adrenal stress responses following septal lesions in the rat. *Neuroendocrinology* *16*: 225.

Selye, H. 1950. "Stress." ACTA, Montreal, Canada.

Singewald, M.L., Langworthy, O.R. and Kouwenhoven, W.B. 1972. "Medical Followup Study of High Voltage Linemen Working in AC Electric Fields." *Trans. Paper T 73 154-2*, presented at IEEE Power Engineering Society Winter Meeting, New York.

Straub, K.D. and Carver, P. 1975. Effects of electromagnetic fields on microsomal ATPase and mitochondrial oxidative phosphorylation. *Ann. N.Y. Acad. Sci.* *247*: 292-300.

Udintsev, N.A. and Moroz, V.V. Response of the pituitary-adrenal system to the action of a variable magnetic field. *Bull. Exp. Biol. Med.* *77*: 641.

CHAPTER 6

PHYSIOLOGICAL AND BEHAVIORAL EFFECTS IN HUMANS

Organisms exposed to ELF electric or magnetic fields may respond with a change in physiological function or behavioral activity. The effects which have been investigated range from electric field-conditioned behavior in fish to effects on biological rhythms in man and include many other effects in man, mammals, birds, amphibians, insects, molds and even bacteria. This diversity in the areas of investigation is complemented by a wide variety in the form of the fields used in these studies. Both AC and DC electric and/or magnetic fields are used at field strengths which vary from the low values typical of the natural environment to values so large they can be found only in specialized laboratory situations.

In the first sections we discuss the reports of studies performed with human subjects.

6.1 PERCEPTION OF LOW LEVEL ELECTRIC AND MAGNETIC FIELDS

Hamer, 1968

Measurements of small changes in reaction times were used to demonstrate sensitivity to horizontal electric fields at frequencies of 12, 6 and 2 Hz. Twenty-seven subjects were exposed to 4 V/m electric fields created by a pair of vertical metal plates.

Reaction times were measured by response to a 1 kHz audio signal. Twenty-four measurements were taken in 11 minutes during the twice-daily test sessions, which lasted a total 16 minutes each, because the fields were on for 5 minutes before measurement was begun. The dual sessions were repeated for 15 days, in which no external fields were applied on the first 5 days and, on the last 10 days, a 4 V/m field was applied for 16 minutes at one frequency and then, after a rest, 16 minutes at another frequency. It was observed that under the influence of the higher of the two frequencies, reaction times were slowed by an average of 1.6 millisecond (standard error, 0.6 millisecond). This result was found when either 12 and 6 Hz or 6 and 2 Hz was used as the pair of test frequencies. Over the course of the 10-day exposure to the fields, there was a steady 5 millisecond increase in the reaction times. In addition, the data show a sharp, 6 millisecond increase in the reaction time coincident with energizing the field on Day 6.

The conclusion that reaction times are increased at the higher frequencies is said to be supported at the 95% level of confidence although the data analysis is difficult to follow. On the basis of this pilot quality study, it appears that human reaction times may be sensitive to external electric fields in the frequency range 2 - 12 Hz and that there may be the ability to discriminate between frequencies in this range.

Konig, 1974b

Reaction times to a visual stimulus (light) were measured for human subjects exposed to a weak electric field at 3 and 10 Hz. The vertical electric field of about 1 - 2 V/m was produced between mesh sheets located in the floor and ceiling of the experimental room. Although too few subjects were studied to allow a statistical analysis of the results, Konig reported that the 3 Hz electric field produced an increase in the reaction times of up to 24 milliseconds, while the 10 Hz field shortened reaction time by 30 milliseconds. The effect was said to persist for at least 40 minutes after the field was turned off.

Konig also found that the 3 Hz fields at 1 V/m and 5 V/m affected the galvanic skin response. At 5 V/m, a change was observed in 5 of 10 subjects.

Konig also reports that his subjects complained of headaches, general fatigue, tightness of the chest and sweating of the palm or hand after several minutes' exposure to the field. After several hours of the test session these complaints disappeared.

Konig's interest in ELF fields stems from his interest in the influence of normal atmospheric ELF disturbances on human behavior. He has demonstrated such an effect in data taken from subjects visiting a 1953 German trade exhibition in Munich. He observed that the reaction times changed in correlation with spontaneous changes in atmospheric ELF fields, and that the increases of about 3 standard deviations from the mean occurred within an hour after the ELF field change, and returned to normal limits within 90 minutes. The duration and intensity of the ELF atmospheric events which occurred in correlation with the reaction time changes are not given.

Friedman et al., 1967

The effect of a magnetic field on human reaction times was investigated at 5 - 11 G at the very low frequencies of 0.1 and 0.2 Hz. While seated in the magnetic field created by two vertical Helmholtz coils adjacent to their heads, the subjects responded to a light by the release of a telegraph key. The data showed statistically significant differences ($p < 0.05$ or 0.10) for subjects exposed to the 0.2 Hz or 0.1 Hz field situations when compared to the control subjects. The 0.2 Hz exposure lengthened the reaction time by about 30 milliseconds -- much more than in the 0.1 Hz case.

The peculiarly low frequencies were chosen for two reasons: a) the observation that the DC potentials in the human and animal cerebrum undergo changes at this rate when there is a change in the level of consciousness; and b) the observation that if external currents are used to cause changes in consciousness, these currents are more effective if modulated at very low frequencies.

Milburn, 1971

This test for possible human perception of electric and magnetic fields required the subject's decision on whether or not a low level electric or magnetic field was turned on. The fiel (stimulus) was presented for 0.5 second randomly in either one of two intervals and the subject responded by determining in which interval the field was present. In this scheme, random responses result in a 50% correct score, while perception with perfect accuracy yields a 100% score. Exposures were made in an acoustically shielded chamber, which incorporated a grounded Faraday cage to isolate the subject from external electric fields and the subjects used earplugs to further decrease the change of audible cues. The electric and magnetic fields were produced between plates or coils surrounding the subject's head while he sat in the chamber, in front of a light panel which indicated the response period. Separate studies utilized magnetic fields of 1 G at 40, 400 and 4,000 Hz or electric fields of 655 V/m at the same frequencies. The subject's response during a 5 secon period was measured 700 times in each experimental session, which lasted over 1 hour. In another study investigating the degree to which the subjects could be trained to perceive the fields, 500 trial blocks were observed.

No field recognition was seen in the 36 subjects exposed to either electric or magnetic fields. Neither of the 2 subjects exposed to electric fields of 655 V/m or 2.6 kV/m at 400 Hz showed any evidence of learning behavior. Learning was seen for one subject, exposed to a 1 G, 4 kHz magnetic field, but 7 other subjects exposed to a similar magnetic field were not able to perceive it even after 14 sessions. The one sensitive subject achieved a score of 61% correct responses.

Further examination of the sensitive subject, who was the experimenter's wife, in magnetic fields of 0.75 - 0.05 G at 4 kHz demonstrated that the number of correct responses varied directly with field intensity. Her sensitivity to fields of different frequencies showed a peak -- with 90% correct responses at about 200 Hz for 1 G exposures. At 0.5 G, she showed no perception of the field.

One other subject showed a sensitivity to a 400 Hz magnetic field that was marginal until, unknown to the experimenter, she took a muscle-relaxing drug, Soma(R) (carisoprodol), to relieve neck pains resulting from a recent accident. After this inadvertent administration of a drug, her score rose to 69% correct, and declined to 60% and then 45% soon thereafter. This second sensitive subject demonstrated learning responses for the 400 Hz magnetic field that were similar to those demonstrated by the other sensitive subject, and her response at low intensities was also similar.

Wieske, 1963

This author gives his personal experience with several individuals unusually sensitive to electric fields. In one case, the subjective recognition of a 60 Hz electric field, apparently without audible cue, was correlated by measurements made with a field meter. The subjects reported that they "heard" the presence of the electric field, though clearly it was not by audible means. Based on the subjects' experiences, Wieske speculates that the nerve endings of the inner ear and skin may be receptor sites. It is not possible to estimate the electric field intensities from the information given.

Schmitt and Tucker, 1973

An experiment to discover if people can perceive 60 Hz fields of approximately 10 G was conducted in an artificial field created by coils surrounding an exposure chamber measuring 2 meters on a side, and a second study was made using coils just large enough to include the subject's head. The field was presented for a short period of time (10 seconds) during which the subject decided if he perceived it or not. The experiment, under control of a minicomputer, ran for about 20 minutes at a session, with the subject sitting in a shock-mounted chair or, alternatively, on a padded non-metallic couch. The chamber was designed with a large degree of acoustical, visual, and vibrational isolation.

Inconclusive data was obtained at 7.5 G using the large coil which exposed the entire person but which, the authors felt, was poorly isolated. Out of 150 trials, the subjects had a grand average of 84.8 correct responses, significant at a level of $p < 0.1$. More remarkable than this questionable result were the cases of certain subjects who were uncannily accurate in perceiving the field-on condition, achieving very small "p" values. Thus, despite the experimenters' best efforts to isolate possible clues (vibration or noise, for example), certain subjects did perceive the fields.

The second experiment used the small coil exposing just the head to a 15 G, 60 Hz field. The average number of correct choices in 150 trials was 75, indicating no perception and in this case there were no outstanding examples of high accuracy.

The authors' analysis of the overall experiment demonstrated a residual "p" value of 0.005, indicating perception. They say, "...very likely it is subliminal vibrations, sound, or other clues, although this has not been proven." They also mention that several subjects reported great certainty that they were sensing the field, although they could not state whether it was something they heard, felt, or otherwise perceived.

Because of the contrast between the data of the small and large coils, the authors feel that there were clues allowing perception in the large coil experiment. An alternative interpretation, of course, is that perception requires more than exposure to the head.

6.2 PERCEPTION, BEHAVIORAL AND PHYSIOLOGICAL RESPONSES IN A MAGNETIC FIELD

Beischer et al., 1973

The details of the 10 and 22.5 hour exposures to a 1 G (45 Hz) magnetic field used in this study on humans were presented in Section 5.2 in connection with the evidence that serum levels of triglycerides were elevated 1 or 2 days after the exposure. The experimental protocol involved many physiological and neurological tests which were negative with the exception of the psychological and behavioral tests analyzed by Gibson and Moroney (1974), Section 6.3.

Specifically, none of the subjects was able to perceive the field and there were no indications of effects on blood pressure, temperature, heart rate, respiration rate or on cardiac function (by electrocardiography). The normal biological rhythms were unchanged. The analysis of respiratory gas showed no effects, as did the tests of cardiovascular, neuromuscular and metabolic functions measured under physical stress produced by use of a pedal ergometer. Reaction times, measured twice daily by 300 trials, were unaffected.

The subject's neurological state was further evaluated by the techniques of pupillography (measuring pupil diameter for a 15-minute period), which is a measure of the state of alertness or drowsiness, and by a test for the critical flicker frequency (measuring the maximum frequency at which a pulsed light source is perceived as a flickering signal). No effects which could be correlated to the magnetic field were observed.

Another measure of overall nervous system function was made using a tracking device in which the subject had to keep a needle centered on scale by response to movements of the needle initiated by an electronic device. Success was measured by the deviations from the center position throughout the 1-minute test period. Motor coordination was examined with the Minnesota Rate of Manipulation Test, which involves turning over 60 blocks in as short a period of time as possible. No magnetic field sensitivity was demonstrated in either of these tests.

6.3 PSYCHOMOTOR FUNCTION AND COGNITIVE SKILLS IN A MAGNETIC FIELD

Gibson and Moroney, 1974

This report concerns data on the psychomotor or cognitive functioning of persons exposed to a 45 Hz 1 G magnetic field in the previous experiment (Beischer *et al.*, 1973). The psychomotor tests, previously described, were the Minnesota Rate of

Manipulation Test and a test involving setting a meter pointer. No significant differences were observed.

However, statistically significant differences were shown in the two cognitive tests, the Wilkinson Addition Test and the Response Analysis Tester. The Wilkinson Addition Test requires that the subject complete correctly as many additions as he can within one minute. The full test consists of 5 one-minute periods. All 6 of the experimental subjects demonstrated decreased performance on the day in which they were exposed to the magnetic field, followed by significantly improved performance after the field was turned off. The authors state: "There seems to be no alternative to the conclusion that the experimental subjects' speed of performing additions decreased during their exposure to an ELF magnetic field." The performance speed was recovered by the time of the first test session in the after-exposure time period.

A Response Analysis Tester (RATER) involves responses made in a decision-making situation requiring short-term memory. The subject must push the unlabeled button which corresponds to the stimulus presented on the screen, at the rate of once every 1½ seconds. Scoring is by the number of correct responses. Short-term memory was measured by requesting the subjects to respond to: a) the symbol presented; b) the previous symbol; c) the symbol before last; and d) the symbol twice removed from the last. Significant experimental results were found with the last, most difficult memory test ("the three-delay mode"). Although the experimental and control groups differed in their scores in the three-delay mode even before the field was turned on, a statistical analysis showed that the experimental group improved more than the control group after the 24 hour magnetic field exposure.

In these experiments, none of the subjects was able to perceive the magnetic field: an analysis of their choices demonstrated that they did not know when the field was on.

The authors' summary stresses that caution be exercised in interpreting the results. Specifically, they say that the lack of a massive experimental effect and the small number of subjects preclude drawing casual relationships between ELF magnetic fields and the few, small changes noticed. They also emphasize that, on balance, few changes were noticed while there were many tests indicating no decrement in performance. Thus, though the effects are clearly demonstrated, their subtlety and rarity require more research.

6.4 EFFECT OF "ZERO" MAGNETIC FIELD

Beischer, 1965

As a corollary to the question of the effect of low level magnetic fields, an experiment was performed to demonstrate possible effects upon the removal of the earth's constant magnetic field of about 0.5 G. Exposure of 6 human subjects to a field attenuated to 1/1000 of the geomagnetic field for a period of 10 days demonstrated an effect on the flicker fusion threshold which decreased in frequency linearly with time during the 10 day test period, so that at the end, the flicker fusion frequency was about 8 Hz compared with the self-control value of about 14 Hz, to which it returned after the test period. The other physiological and psychological parameters measured in this experiment (electrocardiagram, electroencephalogram, blood pressure, respiration rate, and psychophysiological tests) showed no effects. Once again, the low level effect observed required a long period of time to demonstrate itself.

6.5 PHYSIOLOGICAL EFFECTS ON SUBSTATION WORKERS

Sazonova, 1967

The physiological effects of working in a 400 - 500 kV switchyard (50 Hz) are reported here (see Section 5.2 for exposure conditions). The 54 substation workers were divided into two groups according to their working day exposure to electric and magnetic fields: Group 1, exposed an average of 2 hours, consists of 29 "operating" personnel who usually work in a building where the fields are low and may make "rounds" into areas of strong electric fields; and Group 2, exposed an average of 5 hours daily, consists of 25 "maintenance" personnel exposed to electric fields whose strength varies with the task of the day. The workers were tested before and after work for neurological performance and overall well-being, using the physiological parameters of pulse, temperature, blood pressure, and a number of psychomotor tests, including the Critical Flicker Fusion Frequency Test. The tests were conducted on 6 working days in one week. Numerical data for body temperature, pulse and blood pressure readings, taken before and after work for both groups of workers, are presented in Table 6-1a. The "p" values in the chart demonstrate that there are significant differences between the two groups of persons for both the "before work" and "after work" data. Neither group showed much change during the work day except for the pulse rates.

The excitability of nerve and muscle tissues in the hand was evaluated by a test in which the adductor pollicis muscle is stimulated electrically. The test involves measuring the threshold voltage necessary to obtain muscle contraction as a function of the pulse length. Group 2 workers were more easily excited after working day exposures, while Group 1 showed no change during the work day. In another measure of thumb movement at the end of the working day, Group 2 subjects demonstrated a frequency dependent change before and after work, while Group 1 subjects exhibited no change. (See Figure 6-1).

The critical flicker frequency test did not disclose differences between Groups 1 and 2. After-work reaction times for

TABLE 6-1. EFFECTS OF ELECTRIC FIELDS ON PHYSIOLOGICAL TESTS ADMINISTERED TO SOVIET SUBSTATION WORKERS*

a) Indices of Autonomic Function

			Before Squatting	After Squatting	After 5 Minutes of Rest
1. Body Temperature (°C)	Before work	Group I	36.6 ± 0.05	-	-
		II	36.8 ± 0.08	-	-
		Differences	-0.2		
		("P" value)	(0.06)		
	After work	Group I	36.9 ± 0.06	-	-
		II	37.1 ± 0.06	-	-
		Differences	-0.2		
		("P" value)	(0.09)		
2. Pulse Rate (Beats/min)	Before work	Group I	68 ± 2.4	117 ± 2.4	67 ± 1.5
		II	69 ± 2.7	107 ± 5.5	70 ± 2.6
		Differences	-1	+10	-3
		("P" value)	(0.38)	(0.05)	(0.16)
	After work	Group I	73 ± 2.0	115 ± 2.6	75 ± 2.1
		II	67 ± 2.0	98 ± 5.0	69 ± 2.4
		Differences	+6	+17	+6
		("P" value)	(0.002)	(0.002)	(0.04)
3. Systolic Arterial Pressure (mm Hg)	Before work	Group I	121 ± 2.0	140 ± 2.8	119 ± 2.9
		II	107 ± 3.4	126 ± 4.4	101 ± 3.6
		Differences	+14	+14	+18
		("P" value)	(0.001)	(0.004)	(0.001)
	After work	Group I	120 ± 2.5	142 ± 2.9	118 ± 2.8
		II	108 ± 2.0	130 ± 4.7	97 ± 2.6
		Differences	+17*	+12	+21
		("P" value)	(0.001)	(0.01)	(0.001)

* Sazonova, 1967

-continued-

TABLE 6-1. EFFECTS OF ELECTRIC FIELDS ON PHYSIOLOGICAL TESTS ADMINISTERED TO SOVIET SUBSTATION WORKERS

-continued-

a) Indices of Autonomic Function

			Before Squatting	After Squatting	After 5 Minutes of Rest
4. Diastolic Pressure (mm Hg)	Before work	Group I	60 ± 1.9	69 ± 1.8	75 ± 1.9
		II	51 ± 3.5	53 ± 4.1	61 ± 3.9
		Differences ("P" value)	+6 (0.07)	+16 (0.001)	+14 (0.001)
	After work	Group I	60 ± 1.7	61 ± 1.9	72 ± 1.7
		II	52 ± 2.9	50 ± 4.2	55 ± 2.7
		Differences ("P" value)	+8 (0.01)	+11 (0.009)	+17 (0.001)

Note: Group I Low exposure to substation electric fields, less than 2 h daily (29 persons)
Group II High exposure, more than 5 h daily (25 persons)

-continued-

TABLE 6-1. EFFECTS OF ELECTRIC FIELDS ON PHYSIOLOGICAL TESTS ADMINISTERED TO SOVIET SUBSTATION WORKERS

-continued-

b) Indices of Speech-Motor Reactions

Stimulus		Group	Reaction Time (seconds)	Group	Reaction Time (seconds)
1. Bell	Before work	I	0.51 ± 0.08	II	0.49 ± 0.06
	After work	I	0.52 ± 0.15	II	0.53 ± 0.09
	Differences	I	-0.01	II	-0.05
	("p" value)	I	(0.461)		(0.309)
2. Red Light	Before work	I	0.52 ± 0.16	II	0.52 ± 0.19
	After work	I	0.56 ± 0.11	II	0.82 ± 0.11
	Differences	I	-0.04	II	-0.30
	("p" value)	I	(0.421)	II	(0.081)
3. Green Light	Before work	I	0.78 ± 0.19	II	0.76 ± 0.20
	After work	I	1.09 ± 0.14	II	1.31 ± 0.12
	Differences	I	-0.31	II	-0.55
	("p" value)	I	(0.097)	II	(0.009)
			Errors (%)		Errors (%)
4. Green Light	Before work	I	9.1 ± 1.6	II	12.6 ± 2.6
	After work	I	12.4 ± 1.1	II	27.4 ± 3.2
	Differences	I	-3.3	II	-14.8
	("p" value)	I	(0.135)	II	(0.001)
5. Blue-green Light	Before work	I	8.3 ± 1.0	II	14.1 ± 2.3
	After work	I	10.7 ± 2.1	II	32.4 ± 2.8
	Differences	I	-2.4	II	-18.3
	("p" value)	I	(0.159)	II	(0.001)

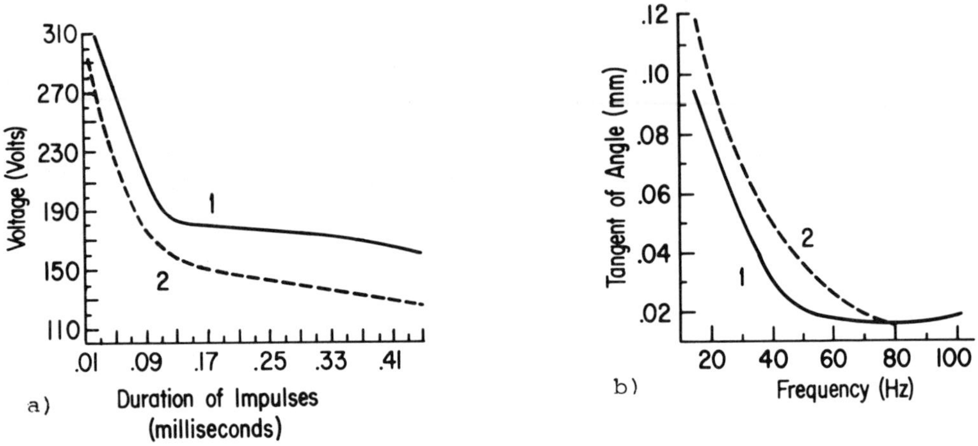

Figure 6-1. Neuromuscular responses of Soviet substation workers

a) Excitability of neuromuscular apparatus of the hand. 1--Before work. 2--After work. Shown for the high exposure group only.*

b) Functional mobility of the hand is shown by the dependence of the tangent of angle of inclination of tetanus curve to initial line on the frequency of the stimulating current. 1--Before work. 2--After work. Shown for the high exposure group only.*

* According to the text, the curves for the other group "coincides with the corresponding dotted lines" in these graphs. Presumably the "dotted line" is actually Curve 1.

Source: Sazonova, 1967

---

visual stimuli were longer for Group 2 (see Table 6-1b) and these workers showed a greater change during the day. An audio stimulus did not have an effect on reaction time.

The author's conclusion is that the "neurodynamic" changes in Group 2 were significantly greater than in Group 1. They also state, "prolonged action of the electric field, evoking inhibitions in the cortex of the brain, favors the development of fatigue."

"The results of physiologic studies agrees with and supplements the data of medical observations. The physiological studies showed vagotonic changes in the cardiovascular system, weakening of the processes of thermal regulation, lowering the resistance to the action of electric currents, and functional mobility of the neuromuscular apparatus of the hand, and the weakening of neurodynamic processes in the cortex of brain. The functional changes recounted above were evident in maintenance personnel (prolonged work in the electric field no less than 5 hours per shift) and they did not appear or appeared weakly in operating personnel (prolonged work in the electric field of not more than 2 hours per shift). This permits us to say that the extent of functional changes in the organism are in direct dependence upon the duration of work under conditions of the action of electric fields..."

Asanova and Rakov, 1966

Medical examinations of the subjects exposed as above (detailed in Section 5.2) disclosed "neural pathology" in 28 subjects. The disorders occurred while the person was in the field and disappeared quickly after they left the field. Subjective complaints were headaches, sleepiness and sluggishness after work, sexual weakness developing after 8 to 10 months of work in substation electric fields, palpitations, chest pains, irritability, and poor appetite.

The medical examination disclosed neural pathology in 28 subjects appearing chiefly as "automonic dysfunction" and less often (6 cases) as "neurasthenic syndrome." Automomic dysfunction was found to be of moderate degree: "stable red dermograph, instability of the pulse and arterial pressure, hyperhydrosis in the armpits and on the palms, fine tremors of the extended hand."

The neurasthenic syndrome found in 4 subjects is described as "hypersthenic" in which the clinical findings were an enhanced tendon reflex, coarse tremor of the fingers, mild instability with closed eyes, and higher emotional excitability.

The medical report continues that no organic damage to the nervous system was seen symptomatically, and it was not possible to establish a relation between complaints and work experience, but they say "more marked neurodynamic changes were seen in maintenance people who had the greatest daily exposure to the electric fields." Apparently, the severity of the findings depended more on the daily exposure than on the number of months of exposure.

The medical findings of organic pathology included incidences of hypertension, arteriosclerotic disease, heart disease, bradycardia (slow pulse) and tachycardia (fast pulse), and hypotension. The severity of these abnormalities is not indicated. Finally, electrocardiograms showed a sinus bradycardia (14 subjects) and slowed atrial and ventricular conductions (10 subjects).

The conclusions drawn are that the instability of various vital signs and EKG changes in persons not having heart diseases are "provisionally considered as functional disorders of the cardiovascular system...evidently connected with disorders of central autonomic regulation." No numbers are given but "the impression is created that changes in the cardiovascular system were encountered more frequently and more markedly in persons systematically subjected to the influence of electric fields..."

Belyaeva and Stroikova, 1975

An abstract of another Soviet report mentions research on 9 human subjects, exposed 2 meters from a 65 kV conductor for periods of 0.5 to 3 hours. The field effects reported were a "slight lowering of body temperature, modification of thermoregulator mechanisms, altered vascular regulatory mechanisms, and lowered work efficiency."

6.6 EFFECTS OF ELECTRIC DISCHARGES (ELECTROSTATIC INDUCTION) IN MAN

Krivova, 1968

A laboratory study of the effects of different electric discharge shocks (secondary shocks) was made on four persons. The experimental conditions are unclear, but apparently were designed to be relevant to exposures near high voltage equipment where, typically, the discharge of a 10 kV potential amounts to a discharge energy of 5 millijoules (mJ), assuming a body capacitance of 100 pF. The pulsed discharges had the effect of increasing the excitability of the brain cortex, depressing the alpha rhythm in the lobar region of the brain, and affecting the "latent period reaction time." Experimental subjects had a shift in the bilateral focus activity of the brain, so that the right rather than the left hemisphere dominated, while this was not observed for the controls.

Effects on autonomic function were also seen in the pulse and respiration rates which changed when subjects were exposed to pulses applied at intervals of from 1 to 30 seconds, but not when the pulses occurred regularly at 20 second intervals. The variability in the pulse and respiration rate was remarkably large, "one-and-a-half to two times" showing a change in the "functional stability" of the organism.

It should be emphasized that here, unlike other reports, the effects are due to a discharge and not the field.

Krivova et al., 1975a

Twenty-three human subjects were exposed to: a) electric fields of from 10 - 32 kV/m; b) body currents equivalent to those induced by an electric field; and c) spark discharges while in the field. No results from study (a) are reported. The results of the current-only experiment (b) indicate that the harmful effects already known from the previous research on substation workers cannot be explained only on the basis of current flow. The third experiment studied changes in pulse and respiratory rates and changes in the amplitude of the electrocardiagram's R-S wave. The response of autonomic function to discharge stimuli below the pain threshold was observed, and the adaptation to pain was found to be greater for regular vs. random rates of discharge. At present, this research is known only through the translator's abstract.

6.7 PHYSIOLOGICAL STUDIES ON DISCHARGE EFFECTS IN MAN AND ANIMALS

Takagi and Muto, 1971

This study examines the physical and mental effects of electrostatic induction which results in secondary discharges. Sensitivity to secondary discharges was examined using an umbrella to typify the situation in which shocks are received from a partially insulated object above ground. They found that perception of static discharges depended upon the manner in which contact was made, the portion of the body to which contact is made (neck and cheek being twice as sensitive as fingers), and no substantial differences exist in the sensitivity of men, women, and children. For the most sensitive situation, they report pain from static discharges occurring in fields above 6 kV/m, and the limits of perception are at about 3 kV/m (ground level field measurements). In these cases the voltages induced in the umbrella are 12 kV and 6 kV, respectively, corresponding to the field intensity.

Tests on humans were also made using: a) induction from an aluminum disc in the electric field near a simulated transmission line; and b) tests made on subjects receiving shocks from a high voltage pulsed DC source. Reactions to these static discharges were measured from electrocardiographs, skin resistance, and pulse rate. The 13 subjects exposed to discharges (0.1 - 0.5 seconds in duration) applied to the forefinger or right cheek while under the influence of a 2.6 kV/m electric showed changes in the galvanic skin response (GSR), the surface resistance of the skin. A gradual increase in the GSR index was observed over a period of several minutes when the discharges were applied at the rate of 1 per minute. Additionally, after a rest period the GSR value declined even more rapidly. Within 5 minutes of application of the first discharge, the GSR declined by 30% of its initial value. No comparison with the control group was presented. The EKG showed no effects due to the electric discharges and the pulse showed a decline of between 3 and 7 per minute (bradycardia). A transient elevation of the blood pressure was observed for subjects exposed to discharges in a field of 2.5 kV/m.

The response of rabbits to electric field discharges was examined, with generally negative results, except under two specific test conditions, which resulted in declines in the blood pressure. Discharges through a gap of 4 mm for 2 seconds caused lowered blood pressure, although discharges through 3 and 5 mm gaps did not. These results for discharges in an electric field could not be reproduced using a pulse simulator.

An attempt to measure conditioned behavior of mice is reported with inadequate statistics. Apparently, mice which took about 5 seconds to travel to food took 9 seconds after being shocked and then 5 seconds after they recovered.

Apart from the difficulties of a poor translation, this report is flawed by a lack of data and statistical analysis of data.

6.8 PHYSIOLOGICAL STUDIES IN MAN

Hauf, 1974

In addition to the hematological measurements reported earlier (Section 5.2), this experiment found there was no influence from electric fields in tests of human reaction time, blood pressure, pulse and in recordings of the electroencephalogram and electrocardiogram. Reaction times to a random signal (5 signals per trial) were measured each 5 minutes during a test session comprised of a 45 minute period of exposure to the field surrounded by two 45 minute control periods. No significant differences were seen in the subject's reaction times during exposure to fields of 1 and 15 kV/m or to the sham field.

Blood pressure measurements (Figure 6-2) indicated no electric field influence during the two 45 minute periods during which fields of 1 or 15 kV/m were turned on. In a similar test, the pulse rate was unaffected by the subject's exposure to the electric field.

The electrocardiagram and electroencephalogram recordings were made during intervals of 5 - 10 seconds during which the field was turned off. Again, the electric field did not affect these measures of heart and brain function.

Kouwenhoven et al., 1967; Singewald et al., 1972

Medical examinations of 10 linemen working on high voltage transmission lines (765, 345 and 138 kV) were conducted over a period of 9 years. Four of the men regularly worked "barehanded" on live lines thus, for example, exposing themselves to 60 Hz electric fields of up to 470 kV/m (at the head) when working on 345 kV lines. These men wore conductive clothing and gloves to shield them from the strong fields (and corona effects). The other 6 employees worked mostly with the "hot stick" technique and were, for example, exposed to maximum fields of 70 kV/m on the 345 kV lines. By the end of the study 8 of the men had taken supervisory jobs which still involved climbing the towers and exposure to electric fields.

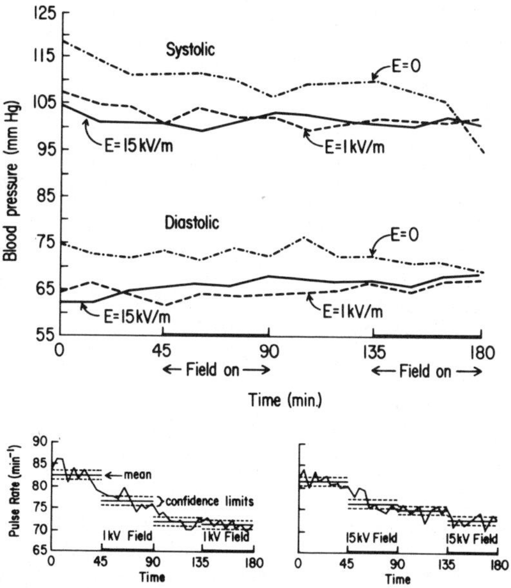

Figure 6-2. Null effects on blood pressure and pulse rate observed in humans exposed at 1 kV/m and 15 kV/m (50 Hz).

Source: Hauf, 1974

The negative findings reported in the studies of blood chemistry and hematology have been mentioned previously (Section 5.2). The other medical examinations made during each 3½ day long examination included electrocardiography, electroencephalography, tests of hearing, vision, X-rays of the chest and hands, and a urological exam including measurements of sperm number and motility. The services of an internist, ophthalmologist, otolaryngologist, urologist, and neuro-psychiatrist were used for each of the 7 exams.

The only abnormal findings observed during the 9 year period were a benign lung lesion in one subject and occasional low sperm counts, which were interpreted as the result of normal variations in sexual habits. The authors conclude that no effects of electric field exposure could be observed.

Johansson et al., 1973

The behavioral responses of human subjects in a 50 Hz electric field of about 20 kV/m were examined in 10 men and 10 women, using psychological tests and questionnaires. The electric field, produced between electrodes in the ceiling and floor, is calculated as 20 kV/m. The exposure room was constructed with a high degree of acoustical isolation. Measurements of the electric field strength were made using a grounded half sphere at various heights above ground. In this way 100 kV/m was determined as the distorted electric field for a subject sitting at a table. The subjects wore insulating shoes (wooden). The measurements with the grounded sphere are apparently values of the distorted field and seem irrelevant. The ion density, humidity, etc. in the room and weather conditions outside were also noted during the experimental period. The psychological tests involved drawing of circles at as high a speed as possible, and reaction times were measured with a standard device in which the subject responds to a light signal with the appropriate key.

The results were generally negative, and none were found to be statistically significant, including a slight decrement in the learning demonstrated on the reaction times tests for the experimental group. The data in the experiment are reported completely but not subjected to as thorough statistical analysis as possible.

Krumpe and Tockman, 1972; 1974

An extensive and well-documented medical survey was made of the personnel employed at the Project Sanguine test facility. Medical examinations were made in two successive years and a statistical analysis was conducted with the data from subjects who could be matched in pairs according to sex and age. Exposures to the magnetic field (45 - 75 Hz) were generally less than 0.1 G and to the electric field less than 0.05 V/m for cumulative periods of up to several thousand hours.

Seventy-eight physiological and laboratory measurements were made and, as reported earlier (Section 5.2), the blood chemistry and hematology findings were negative. In this study, the neurological and physiological problems found in Soviet substation workers by Asanova and Rakov (1966) were given special attention. No effects were seen in endocrine gland function, in electrocardiograms, electroencephalograms , measurements of nerve conduction velocity, audiograms, ophthalmological tests and in psychological and emotional tests.

6.9 EFFECTS ON BIOLOGICAL (CIRCADIAN) RHYTHMS

Many body functions -- temperature, hormone levels, alertness, etc. -- vary periodically throughout the day, with a period of about 24 hours, but the means by which organisms, including man, maintain this regularity is not yet understood. Recently the hypothesis that the circadian "clock" results from interactions between the organism and the electromagnetic environment have been given increasing attention (e.g., Brown, 1972).

Wever, 1967; 1968; 1974

The circadian rhythms of human subjects were investigated during long term exposures to an external electrical field of 2.5 V/m at 10 Hz (square wave). Experiments lasting 3 - 4 weeks in an isolated chamber involved two factors: the influence of natural electromagnetic fields determined by comparison of subjects in the shielded and non-shielded rooms; and the effect of an artificial electric field (10 Hz) introduced into the shielded room. The investigator measured the times of waking and sleeping, degree of activity and motion, body temperature, electrolyte activity (by urinalysis), and administered several tests. The subjects exhibited no perception of the applied field, but field did not have the effect of shortening the circadian period by 1 hour after 9 days of continuous exposure to the field. When the field was turned off, the period then lengthened by 1.27 hours (0.23 hour standard error) which is a significant change ($p < 0.001$). Another indication of a field effect is that while the field was on there was a considerably greater ratio of activity time to resting time ($p < 0.05$). (See Table 6-2.)

A portion of this study demonstrated that subjects in a room screened from natural electromagnetic fields had a slightly longer period, also statistically significant ($p < 0.001$). Wever notes that in this case too the field-free time period had the longer circadian rhythm compared with times when either a natural or artificial field was present. Thus, Wever feels that external natural or manmade fields accelerate the normal periodic rhythms in man. The last portion of this experiment investigated the phenomenon of desynchronization in which the activity cycle is no longer in step with the physiological cycle. It appears that either the external field or the natural field in the unscreened room inhibits desynchronization.

A preliminary investigation of DC electric and magnetic fields indicates that weak (300 V/m, and 1.5 G) homogeneous DC fields do not affect biological rhythms. These negative results lend greater credibility to the 10 Hz findings discussed above.

6.10 CORRELATION OF HUMAN BEHAVIOR AND GEOMAGNETIC STORMS

Friedman and Becker, 1963

A statistical correlation was found between admissions to a psychiatric hospital and changes in the geomagnetic field. Data were gathered from 7 hospitals in the same area over a period of 4 years and correlated with geomagnetic data taken by the Coast and Geodetic Survey. The authors comment upon the well known fact that correlation does not imply causal relationships and conclude with their impression that: "In view of the diversity of the two variables under investigation, and the grossness of the measures, the present investigations offer correlations of a surprising magnitude."

6.11 MISCELLANEOUS REPORTS

Tromp, 1949

This encyclopedic volume describes numerous investigations into the influence of the natural and artificial electric and magnetic fields. The influences of weather and the degree of air ionization are considered among other environmental factors.

There is a description of the clinical use of static electricity (p. 182) in which persons are treated for periods of 20 minutes or so, at voltages sufficient to make the hair stand on end. Beneficial medical effects were claimed for these

TABLE 6-2. EFFECT OF A 10 Hz (SQUARE WAVE) ELECTRIC FIELD ON THE BIOLOGICAL RHYTHM IN HUMANS*

Subject No.	Duration of Circadian Cycle Without Field		Duration of Circadian Cycle With Field		Difference in Duration (w/o Field mi With Field)
28	28.2 ± 1.7 hrs	(9)	25.7 ± 1.1 hrs	(9)	2.5 hrs
31*	25.2 ± 1.7 hrs	(12)	24.3 ± 1.6 hrs	(10)	0.9 hrs
33	28.1 ± 6.3 hrs	(13)	26.3 ± 2.7 hrs	(7)	1.8 hrs
35	26.6 ± 1.0 hrs	(6)	25.8 ± 2.6 hrs	(9)	0.8 hrs
39	25.7 ± 4.6 hrs	(10)	24.7 ± 4.2 hrs	(10)	1.0 hrs
41	25.3 ± 1.4 hrs	(11)	24.7 ± 1.4 hrs	(7)	0.6 hrs
44*	25.2 ± 0.8 hrs	(7)	24.8 ± 1.4 hrs	(7)	0.4 hrs
54	25.7 ± 2.1 hrs	(7)	24.9 ± 1.0 hrs	(7)	0.8 hrs
56	27.6 ± 3.2 hrs	(12)	25.3 ± 1.6 hrs	(8)	2.3 hrs
61	26.0 ± 1.0 hrs	(8)	24.4 ± 1.4 hrs	(6)	1.6 hrs

Average Values Standard Deviation (Average Error)

	Without Field	With Field	Difference
	26.36 hrs ± 1.20 (0.38)	25.09 hrs ± 0.66 (0.21)	1.27 hrs ± 0. (0.2

Statistical tests for difference greater than zero

1. t-test: $t = 5.52$
   $p < 0.001$

2. Wilcoxon test: $t = 0$
   $p < 0.005$

In the tests designated "*" measurements were made at first with and then without a field. The cycle values are followed by the standard deviation and the number of cycles measured in parentheses.

*Wever, 1967; 1968

treatments whose efficacy is related to possible changes in electric skin potential.

In another place (p. 255) there is reference to experiments on animals indicating higher rates of cancer induction for mice in grounded cages as compared with mice in cages insulated from ground.

Many other references to biological experiments involving electricity or magnetism are given but these experiments, quite old and often quite curious, are of limited current scientific value.

6.12 DOWSING: A MAGNETIC PERCEPTION?

Tromp, 1968; 1949

Experiments on the phenomenon of dowsing imply links between this curious human activity and the presence of geomagnetic anomalies. One thinks of a dowser holding a forked rod, walking over land to determine where underground water is. However, the dowsing phenomenon is not limited to the determination of water, nor does it seem to require a special rod. The dowsing reaction, expressed as a turning of the extended forearms, can occur under diverse circumstances when the arms are extended with or without a rod.

Tromp (1968) remarks: "It is doubtful whether as much investigation and discussion have been bestowed on any other subject with such a lack of positive evidence for the reality of the observed fact." If substantiated, dowsing is related to our interest in the perception of magnetic fields. Experiments which demonstrate the physiological changes which occur when dowsers enter the "dowsing zone" show marked changes in skin potential, and in the DC level of the electrocardiogram. Using soil resistivity as a measure of subsoil geological discontinuities, Tromp demonstrates that dowsing zones correlate with geologic discontinuities indicated by regions of lowered soil resistivity. His laboratory experiments found that some dowsers could sense magnetic field gradients of less than 0.1 G/m.

6.13 REFERENCES

Asanova, T.P. and Rakov, A.I. 1966. The state of health of persons working in electric field of outdoor 400 and 500 kV switchyards. Hygiene of Labor and Professional Diseases 5. Translation by G. Knickerbocker available in Special Pub. #10, IEEE Power Engineering Society, Piscataway, N.J. (1975).

Beischer, D.E. 1965. Biomagnetics. Ann. N.Y. Acad. Sci. 134: 454.

Beischer, D.E., Grissett, J.D. and Mitchell, R.E. 1973. "Exposure of Man to Magnetic Fields Alternating at Extremely Low Frequency." Naval Aerospace Medical Research Laboratory, Pensacola, Fla. Rept. NAMRL-1180. Available from NTIS as AD 770140 (1973).†

Belyaeva, P.I. and Stroikova, K.V. 1975. "The Physiological Investigation of the Effects of a High Tension, Industrial Frequency Electrical Field on Living Organisms." Abstract by G. Knickerbocker available in Special Publ. #10, IEEE Power Engineering Society, Piscataway, N.J. (1975).

Brown, F.A. 1972. The "clocks" timing biological rhythms. Amer. Sci. 60: 756.

Friedman, H. and Becker, R.O. 1963. Geomagnetic parameters and psychiatric hospital admissions. Nature 200: 626.

Friedman, H., Becker, R.O. and Bachman, C.H. 1967. Effect of magnetic fields on reaction time performance. Nature 213: 949.

Gibson, R.S. and Moroney, W.F. 1974. "The Effect of Extremely Low Frequency Magentic Fields on Human Performance: a Preliminary Study." Naval Aerospace Medical Research Laboratory, Pensacola, Fla. NAMRL-1195.†

Hamer, J.R. 1968. Effects of low level, low frequency electric fields on human reaction time. Comm. Behav. Biol. Pt. A, 2: 217.

Hauf, R. 1974. "Wirkung von 50-Hz-Wechselfeldern auf den Menschen (Effects of 50 Hz alternating fields on man). Sonderdruck aus Elektrotechnische Zeitung-b 26: 381.

Johansson, R. et al. 1973. Is there a connection between the electricity in the atmosphere and the function of man? Part 3, 50 Hz field variations. Forvarets Forskningsanstalt Advelning 2: FOA Report C 2627-H5.

Konig, H.L. 1974b. "Behavioral Changes in Human Subjects Associated with ELF Electric Fields." IN: "ELF and VLF Electromagnetic Field Effects." M. Persinger, ed. Plenum Press, New York, p. 81.

Kouwenhoven, W.B. et al. 1967. Medical evaluation of man working in AC electric fields. IEEE Trans. Pow. App. & Sys. PAS-86, No. 4.

Krivova, T.I. 1968. "Electrical Discharge Influence upon a Man." Proc. 3rd All-Union Symp., Hygiene of Labor and Biological Effects of Radiofrequency Electromagnetic Waves, Moscow. Translation by G. Knickerbocker available in Special Publ. #10, IEEE Power Engineering Society, Piscataway, N.J. (1975).

Krivova, T.I., Lukovkin, V.V., and Sazonova, T.E. 1975a. "The Influence of an Electric Field of Commercial Frequency and Discharges on a Human Organism." Abstract by G. Knickerbocker available in Special Publ. #10, IEEE Power Engineering Society, Piscataway, N.J. (1975).

Krumpe, P.E. and Tockman, M.S. 1972. "Evaluation of the Health of Personnel Working near Project Sanguine Beta Test Facility from 1971 to 1972." Naval Medical Research Unit No. 4, Great Lakes, Ill.†

Krumpe, P.E. and Tockman, M.S. 1974. "Evaluation of the Health of Personnel Working near Project Sanguine Beta Test Facility from 1971 to 1972." IN: "Biologic and Clinical Effects of Low Frequency Magnetic and Electric Fields." J. Llaurado et al. eds. Charles C. Thomas, Springfield, Ill.

Milburn, B.B. 1971. "Very Low Frequency Electromagnetic Fields and Behavior." Dissertation, U.C.L.A., available as #72-2866 from University Microfilms, Ann Arbor, Mich.

Sazonova, T.E. 1967. "Physiological and Hygienic Assessment of Labour Conditions at 400 - 500 kV Outdoor Switchyards." Scientific Pubs., Inst. of Labour Protection of the All-Union Central Council of Trade Unions, Issue 46. Profizdat. Translation by G. Knickerbocker available in Special Publ. #10 IEEE Power Engineering Society, Piscataway, N.J. (1975).

Schmitt, O.H. and Tucker, R.D. 1973. Human perception of moderate strength low frequency magnetic fields. IEEE Electromagnetic Symp. Record, June, 1973, pp. 65-70.

Singewald, M.L., Langworthy, O.R. and Kouwenhoven, W.B. 1972. "Medical Follow-up Study of High Voltage Linemen Working in AC Electric Fields." Trans. Paper T 73 154-2, presented at the IEEE Power Engineering Society Winter Meeting, New York, 1973.

Takagi, T. and Muto, T. 1971. "Influences upon Human Bodies and Animals of Electrostatic Induction Caused by 500 kV Transmission Lines." Tokyo Electric Power Co., Inc. Tokyo, Japan.

Tromp, S.W. 1949. "Psychical Physics." Elsevier Publishing Co., Inc., New York.

Tromp, S.W. 1968. Review of the possible physiological causes of dowsing. _Int. J. Parapsychology 10_: 363.

Wever, R. 1967. "Ueber die Beeinflussung der circadiane Periodik des Menschen durch schwache elektromagnetische Felder (The influence of weak electromagnetic fields on the circadian rhythm in man). _Z.F. Vergleichende Physiologie 56_: 111. NASA translation _X68-13292_.

Wever, R. 1968. Enfluss schwacher elektromagnetischer Felder auf die circadiane Periodik des Menschen (Influence of weak electromagnetic fields in the carcadian periodicity of man). _Naturwissenschaften 55_: 29. NASA translation _X68-18216_.

Wever, R. 1974. "ELF Effects on Human Circadian Rhythms." IN: "ELF and VLF Electromagnetic Field Effects." M. Persinger, ed. Plenum Press, New York.

Wieske, C.W. 1963. Human sensitivity to electric fields. _Biomed. Sci. Instrum. 1_: 467.

CHAPTER 7

PHYSIOLOGICAL AND BEHAVIORAL EFFECTS IN NON-HUMAN SPECIES

The behavior of animals exposed to electric and magnetic fields at ELF has been studied by scientists with interests in brain rhythms, migratory and navigational behavior of birds and fish, and the growth and development of embryos or young animals. Experiments have also been performed observing behavior of animals trained for conditioned response to a stimulus. The motivations for these different experiments have varied. Some, being a part of the Sanguine Project, attempt to determine the ecological impact of the ELF transmitter system; some are the outgrowth of research into brain rhythms; some are concerned with navigational systems employed by migratory birds or fish; and others are from researchers interested in behavioral responses to natural ELF fields, perhaps useful as subtle biological clocks, etc. The principal studies in this broad area are discussed below, beginning with effects on the behavior of monkeys.

7.1 CENTRAL NERVOUS SYSTEM EFFECTS PRODUCED IN MONKEYS BY ELECTRIC FIELDS

Gavalas-Medici and Magdaleno, 1975

The responses of 5 female monkeys to ELF fields (at 7, 45, 60, and 75 Hz with field strengths between 1 - 100 V/M) were examined in a coordinated series of experiments examining behavioral and neurophysiological performance. EEG electrodes were implanted surgically in 3 animals and 2 of the animals, although exposed to the electric fields, did not have electrode implants. In the behavioral experiment the monkeys were trained to press a lever at the correct time to receive a reinforcing squirt of apple juice (technically described as a "fixed interval schedule of reinforcement with differential reinforcement of low rates and a limited hold"). When their behavior was stable enough (70% correct responses) exposure to an electric field produced between parallel plates was begun. Animals were exposed for a total of 4 hours and data were taken during the last 3 hours of the sessions. The datum of the behavioral experiment is the interresponse time (IRT) between lever pressings.

The EEG data were taken only during the last hour of the experimental session and from but 2 animals exposed at the lowest voltage levels and 1 animal at the higher voltage levels. The EEG data which were used for analysis included a representative number of correct and incorrect responses, where the incorrect responses included an even number of fast or slow presses.

At an electric field strength of about 1 V/m, no significant differences were found among the IRT's at any of the 3 frequencies (7, 45, and 75 Hz).

At 10 V/m, there was a suggestion of sensitivity to the 7 Hz electric field in that while the 7 Hz field was on, the IRTs were shorter than for the other field conditions, and the standard deviations were smaller. The findings at 7 Hz are believed to be significant because of the small "p" value and because it is unlikely that the 7 Hz data would be in the first rank (shortest) by chance. There was no difference in the percentage of correct responses or in the total number of lever presses at these field conditions.

Four monkeys, exposed to 56 V/m electric fields at 7, 60, and 75 Hz, exhibited more marked behavioral change. At 7 Hz, the IRT is shortened by 0.29 second, compared with the control level, significant at the 0.05 level, and more than twice the difference found at 10 V/m (7 Hz). Again, the standard deviations in the IRTs were smaller, significant at the 0.01 level. The data at 75 Hz show a significant difference in the mean IRT of 0.29 second ($p < 0.025$), while at 60 Hz, there was no significant difference.

It is possible that the negative 60 Hz, 56 V/m results might be attributed to habituation of the animals to the ambient 60 Hz fields in the laboratory. However, measurements in the animal living quarters revealed a level of only about 0.1 V/m. Thus, the alternative hypothesis, that effects are frequency-selective, is more likely.

The experiment was repeated again at a field level of 100 V/m, at 75, 60, and 7 Hz. Significant findings at all 3 frequencies are found <u>if</u> the "field-on" data are compared with "field-off" data taken at the conclusion of the experiment rather than "field-off" data taken between experimental exposures. If the comparison is made between "field-on" data and the interspersed "field-off" data, no significant effects are found. This peculiarity is explained by the hypothesis that at this voltage level the electric field effects last past the exposure session, thus causing shorter IRT's on the following day, so that a proper control period must be after the animals have recovered from the field effects. However, the "p" values found in the 100 V/m experiment are quite large, varying between 0.1 and 0.2, when using the after-exposure control sessions. The authors comment that their after-exposure sessions may still have been affected by the lingering field effects, explaining the shortened IRTs found in the control data.

The EEG data reported from these experiments show no dramatic effects. Due to the small number of animals for whom EEGs were analyzed, a descriptive report of the findings was given discussing the relative amount of brain wave energy at different frequencies across the EEG spectrum, and possible influences from the imposed electric fields operating at other frequencies. No strong conclusions were drawn from the examination of the EEG data, although some suggestions for further experimentation are presented. (See Tables 7-1, 7-2, 7-3, and Figure 7-1.)

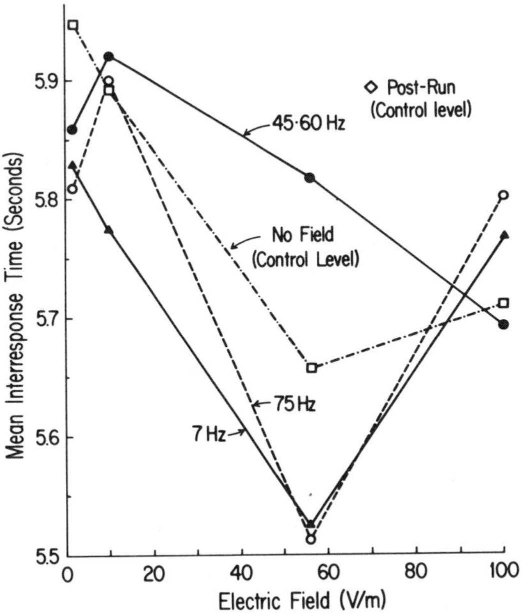

Figure 7-1. Conditioned behavior in monkeys: mean interresponse times vs. horizontal electric field strength at several frequencies.

Source: Gavalas-Medici and Magdaleno, 1975

This experiment demonstrates two important effects which, if confirmed, begin to quantify ELF effects. Apparently, ELF field effect thresholds are frequency-dependent, with lower thresholds at lower frequencies and ELF fields affect the central nervous system in a manner that takes days to fully attenuate. It is suggested that the threshold for ELF effects is in the range of 10 - 50 V/m.

In addition, the authors comment that there were no signs of physical "malaise or gross disorientation at any of the voltage levels tested." They also suggest that "the detection of weak electric field effects in laboratory primates depends on four major variables; 1) frequency; 2) voltage; 3) duration of exposure; and 4) exact behavioral assay being used."

The authors propose that all organisms which exhibit slow electrical waves (similar to the low frequency EEG waves) and which possess membrane surfaces with macromolecular sensors capable of transducing components of the extra-cellular field by an alteration of the ion bonding (by conformational change) may be susceptible to low level electric fields. The changes seen here at several frequencies and field strengths are restricted to small alterations in the interval between monkey responses during a learned task.

TABLE 7-1. CONDITIONED BEHAVIOR IN MONKEYS EXPOSED TO LOW LEVEL ELECTRIC FIELDS AT SEVERAL FREQUENCIES AND FIELD STRENGTHS *

a) Mean interresponse Times and Differences (Field Minus No-Field) for Monkeys Exposed to a Horizontal Electric Field (7 Hz)

Field Cond.	Monkey: A	B	C	D	E	Average Difference
7/1	6.40 s	5.61 s	6.15 s	5.52 s	5.47 s	-.070 s
0/0	6.58	5.87	6.03	5.58	5.44	
diff	-.18	-.26	.12	-.06	.03	
7/10	6.36	5.72	6.03	5.37	5.43	-.118
0/0	6.58	5.87	6.03	5.58	5.44	
diff	-.22	-.15	0	-.21	-.01	
7/56	5.98	--	5.35	5.43	5.30	-.288
0/0	6.08	--	5.96	5.50	5.67	
diff	-.10	--	-.61	-.07	-.37	
7/100	6.52	--	5.68	5.50	5.37	.060
0/0	6.33	--	5.45	5.52	5.53	
diff	.19	--	.23	-.02	-.16	
7/100	6.52	--	5.68	5.50	5.37	-.123
PRO/0	6.49	--	6.05	5.51	5.51	
diff	.03	--	-.37	-.01	-.14	

Legend

7/1 - 7 Hz/1 V/m
7/10 - 7 Hz/10 V/m
etc.
0/0 - Self control in no field

* Gavalas-Medici and Magdaleno, 1975

TABLE 7-1. CONDITIONED BEHAVIOR IN MONKEYS EXPOSED TO LOW LEVEL ELECTRIC FIELDS AT SEVERAL FREQUENCIES AND FIELD STRENGTHS (Cont'd.)

b) Mean Interresponse Times and Differences (Field Minus No-Field) at 45 and 60 Hz

Field Cond.	Monkey: A	B	C**	D	E**	Average Difference
45/1	6.16 s	5.93 s	6.24 s	5.51 s	5.47 s†	-.055 s (45 Hz, A,B,C,D)
0/0	6.58	5.87	6.03	5.58	5.44 †	
diff	-.42	.06	.21	-.07	.03 †	
45/10	6.54	5.72	6.15	5.43	5.74 †	-.055 (45 Hz, A,B,C,D)
0/0	6.58	5.87	6.03	5.58	5.44 †	
diff	-.04	-.15	.12	-.15	.30 †	
60/56	6.06	--	5.76	5.63	5.77	.003 (60 Hz)
0/0	6.08	--	5.96	5.50	5.67	
diff	-.02	--	-.20	.13	.10	
60/100	6.49	--	5.60	5.52	5.13	-.023 (60 Hz)
0/0	6.33	--	5.45	5.52	5.53	
diff	.16	--	.15	.00	-.40	
60/100	6.49	--	5.60	5.52	5.13	-.205 (60 Hz)
PRO/0	6.49	--	6.05	5.51	5.51	
diff	.00	--	-.45	.01	-.38	

**No implanted electrode

† 60 Hz

TABLE 7-1. CONDITIONED BEHAVIOR IN MONKEYS EXPOSED TO LOW LEVEL ELECTRIC FIELDS AT SEVERAL FREQUENCIES AND FIELD STRENGTHS (Cont'd.)

c) Mean Interresponse Times and Differences (Field Minus No-Field) at 75 Hz

Field Cond.	Monkey:	A	B	C	D	E	Average Difference
75/1		6.47 s	5.74 s	6.17 s	5.43 s	5.22 s	-.094 s
0/0		6.58	5.87	6.03	5.58	5.44	
diff		-.11	-.13	.14	-.15	-.22	
75/10		6.40	5.79	6.20	5.61	5.52	0
0/0		6.58	5.87	6.03	5.58	5.44	
diff		-.18	-.08	.17	.03	.06	
75/56		5.69	--	5.58	5.42	5.36	-.29
0/0		6.08	--	5.96	5.50	5.67	
diff		-.39	--	-.38	-.08	-.31	
75/100		6.32	--	5.78	5.50	5.58	.088
0/0		6.33	--	5.45	5.52	5.53	
diff		-.01	--	.33	-.02	.05	
75/100		6.32	--	5.78	5.50	5.58	.095
PRO/0		6.49	--	6.05	5.51	5.51	
diff		-.17	--	-.27	-.01	.07	

TABLE 7-2. CONDITIONED BEHAVIOR IN MONKEYS EXPOSED TO ELECTRIC FIELDS *

a) Mean Interresponse Time and Standard Deviation for Exposure of 5 Monkeys to a Horizontal 10 V/m Electric Field (in seconds)

MONKEY

A			B			C**			D			E**		
Field Cond.	$\bar{x}$IRT	σ	Field Cond.	$\bar{x}$IRT	σ	Field Cond.	$\bar{x}$IRT	σ	Field Cond.	$\bar{x}$IRT	σ	Field Cond.	$\bar{x}$IRT	σ
7/10	6.36	1.73	45/10	5.72	1.28	0/0	6.03	1.13	7/10	5.37	.89	7/10	5.43	1.13
75/10	6.40	1.79	7/10	5.72	1.30	7/10	6.03	1.23	45/10	5.43	.97	0/0	5.44	1.26
45/10	6.54	2.04	75/10	5.79	1.33	45/10	6.15	1.30	0/0	5.58	.98	75/10	5.52	1.28
0/0	6.58	2.06	0/0	5.87	1.66	75/10	6.20	1.33	75/10	5.61	1.09	60/10	5.74	1.42
Field $\bar{x}$	6.43	1.85	Field $\bar{x}$	5.74	1.42	Field $\bar{x}$	6.13	1.25	Field $\bar{x}$	5.47	.95	Field $\bar{x}$	5.56	.95

* Gavalas-Medici and Magdaleno, 1975

**Indicates no EEG leads were implanted in these animals.

The data are in rank order. "σ" is the standard deviation.
The notation of the field conditions is, 7/10 - 7 Hz, 10 V/m; 45/10 - 45 Hz, 10 V/m; 75/10 - 75 Hz, 10 V/m; 0/0 - control, no field

TABLE 7-2. CONDITIONED BEHAVIOR IN MONKEYS EXPOSED TO ELECTRIC FIELDS (Cont'd.)

b) Mean Interresponse Time and Standard Deviation-56 V/m (in seconds)

MONKEY	A		C**			D			E**		
Field Cond.	$\bar{x}$IRT	σ	Field Cond.	$\bar{x}$IRT	σ	Field Cond.	$\bar{x}$IRT	σ	Field Cond.	$\bar{x}$IRT	σ
75/56	5.69	1.18	7/56	5.35	.80	75/56	5.42	.82	7/56	5.30	1.31
7/56	5.98	1.48	75/56	5.58	.90	7/56	5.43	.91	75/56	5.36	1.44
60/56	6.06	1.59	60/56	5.76	.91	0/0	5.50	1.05	0/0	5.67	1.46
0/0	6.08	1.60	0/0	5.96	1.02	60/56	5.63	1.06	60/56	5.77	1.55
Field $\bar{x}$	5.91	1.42	Field $\bar{x}$	5.56	.87	Field $\bar{x}$	5.49	.93	Field $\bar{x}$	5.48	1.40

TABLE 7-3. CONDITIONED BEHAVIOR IN MONKEYS EXPOSED TO LOW LEVEL ELECTRIC FIELDS AT SEVERAL FREQUENCIES AND FIELD STRENGTHS: STATISTICS FOR MONKEY INTERRESPONSE TIMES *

Δ IRT = Field IRT minus No-Field IRT (seconds)
$\Delta\sigma$ = Field std. deviation minus No-Field std. deviation (s)

	1 V/m		10 V/m		56 V/m		100 V/m		PR-100 V/m**	
	Δ IRT	$\Delta\sigma$	Δ IRT	$\Delta\sigma$	Δ IRT	$\Delta\sigma$	Δ IRT	$\Delta\sigma$	Δ IRT	$\Delta\sigma$
7 Hz Size of diff.	-.070	-.012	-.118	-.160	-.288	-.178	.060	-.063	-.123	-.098
p	.20	.50	.05	.025	.05	.01	.30	.35	.15	.10
45 Hz Size of diff.	-.055	.028	-.055	-.010	-	-	-	-	-	-
p	.40	.45	.25	.45	-	-	-	-	-	-
60 Hz Size of diff.	-	-	-	-	.003	-.058	-.023	-.153	-.205	-.178
p	-	-	-	-	.50	.10	.45	.20	.10	.10
75 Hz Size of diff.	-.094	.020	0	.012	-.290	-.225	.088	-.078	-.095	-.113
p	.15	.40	-	.50	.025	.025	.20	.20	.20	.20

("p" measures the statistical significance of the difference in the mean values, e.g. a "p-value" of .05 means that in 1 of 20 trials the difference in the numbers would be expected to occur by chance)

**The "PR-100 V/m" data is obtained using post-run control data due to an apparent residual effect of the fields which made the between-runs control data less useful

* Gavalas-Medici and Magdaleno, 1975

Gavalas et al., 1970

This experiment, utilizing a protocol similar to the one above, discovered that the interresponse time (IRT) was considerably shorter when the monkeys were exposed to a 7 Hz electric field of 7 V/m strength. Electric fields of 10 Hz at the same strength were not effective in changing the animals' behavioral response. In the 7 Hz experiment, the IRT was shortened by 0.4 second or more, and data were highly significant ($p < 0.01$ or < 0.001).

In this experiment, the analysis of the EEG data showed increases in the power in the brain rhythms at the same frequencies as were imposed externally, or at harmonics of the external fields. At the time, it was not clear whether these EEG data were artifacts because the electrodes and leads may act as an antenna introducing energy to the brain causing the supposed field effects. In a later experiment discussed above (Gavalas-Medici and Magdaleno, 1975), only some of the animals had implanted electrodes; yet the implanted and non-implanted monkeys showed the same effects, thus indicating that biological effects of the fields and not induction through implanted leads are responsible in both cases.

7.2 CENTRAL NERVOUS SYSTEM EFFECTS IN CATS USING AN ELF MODULATED RADIO FREQUENCY CARRIER WAVE

Bawin et al., 1973

Research from the same laboratory group in which Gavalas-Medici and Magdaleno work was performed on cats using a VHF (147 MHz) electromagnetic field which was modulated at a frequency between 1 and 25 Hz. The fields were applied between 2 aluminum plates to cats which had been trained to produce a specific "transient brain rhythm" in response to a periodic light flash stimulus. The energy density of the VHF carrier field was 1 mW/cm^2 or less and the experiment demonstrated that the effects observed depended upon the modulation frequency and not the carrier. The use of a VHF carrier was suggested to the researchers by the experiences of microwave irradiations with pulsed wave forms, which implied that the pulse rate was a very significant parameter in determining behavioral effects.

The animals exposed to the modulated VHF signal were exceptionally resistant to a loss of their trained behavior (extinction) and an analysis of the EEGs showed greater regularity of their response to the stimulus with a sharpened spectral distribution of the EEG patterns. In a second experiment, the external field was used to reinforce certain brain rhythms, and it was found that this reinforcement is dependent upon the modulation frequency.

The possibility of an artifact due to an antenna effect is felt to be remote because of the experimental conditions and because the EEG changes reported depended upon anatomical location and were highly specific in frequency and time of appearance.

The authors' discussion makes several suggestions of how such low level effects could have biological effectiveness on the central nervous system, including a hypothesis of how the "greater membrane of brain cells" can be affected by conformational changes of the surrounding coat of proteins due to an interaction with external cations in the intercellular fluid.

7.3 ELECTRIC FIELDS IN BRAIN TISSUE

What are the internal electric fields acting on the neural tissue in the brain at the external field levels used in these experiments? Adey (1975) estimates that for an external field of 10 V/m, the expected electric gradient across the brain tissue itself would be between 10^{-5} and 10^{-6} V/m (in a monkey brain with assumed resistivity of 3 ohm-meters and a maximum dimension of 0.07 meter). These numbers are felt to characterize the internal electric fields at the threshold for behavioral and neurological effects suggested by the experiments of Gavalas-Medici and Magdaleno. Adey's interpretation of behavioral, neurophysiological, and neurobiochemical experiments performed in his laboratory lends support to the hypothesis that "cortical neurons are sensitive to the natural electric field gradients that surround them," as an explanation of both the natural EEG rhythms and the effects observed in the experiments where external fields were applied.

The production of the low frequency brain rhythms observed by EEG techniques -- still not understood -- implies that neurons communicate with each other by the production of a volume-conducted field which is at once a product of the mass of neurons and influences each neuron. Thus, if nerve cells in the brain are sensitive to the neural fields associated with low frequency brain rhythms, they may also be as sensitive to neural fields produced by external influences. A quantification of these ideas in terms of the electric fields surrounding the neurons due to the brain rhythm fields compared with the electric fields due to external fields is not given.

Adey discusses the "greater membrane model" to explain how membrane amplification can occur on the membrane's outer surface, which is composed of a complicated sheet of negatively charged macromolecules electrically balanced by a counterion layer formed of Ca^{++}, protons, and other ions. It is suggested that small variations in the electric fields might result in disturbances of this ionic layer through the mechanism of conformational changes which require very little energy. If the effect is propagated over the membrane, then it may be able to trigger a response in the neuron.

7.4 BEHAVIORAL AND PHYSIOLOGICAL STUDIES OF MONKEYS IN MAGNETIC AND ELECTRIC FIELDS

Grissett and deLorge, 1971; deLorge, 1974

A series of investigations of the effects of magnetic fields on the behavior of squirrel or rhesus monkeys was completed with essentially negative results. Experiments using combined electric and magnetic fields are described below.

Three squirrel monkeys were trained to hold a lever down for 5 seconds and then, in response to a light, release it during a limited hold period of 3 seconds. For a correct performance they received a reinforcing pellet of food. The animals' behavior was observed in 37 daily sessions of 1 hour each: 12 control days of tests; 5 tests in a 45 Hz (3 G) magnetic field; 5 tests in a 7 Hz (3 G) magnetic field; and 15 more test-days with no fields. The response datum was the speed with which the animals released the lever after a light went on, of the order of 1½ - 2 seconds, and the statistics of the number of correct responses, etc. No magnetic field-related changes in behavior were observed upon statistical analysis.

For comparison with the experiments reported above, and those to follow, it should be noted that a stimulus-response time, not an interresponse time or a reaction time, was measured here. The interresponse time task requires estimation of a time interval, whereas reaction time measures a neuromuscular reflex.

Following the change in serum triglycerides observed in an experiment with humans (Beischer *et al*., 1973), another primate experiment with magnetic fields was performed (deLorge, 1974), with assays for both behavioral and blood chemistry changes. The blood chemistry results, which were reported earlier in Section 5.2, were negative. The experiments on 4 rhesus monkeys were conducted during 2-hour exposures at 45 and 15 Hz, with magnetic field intensities of about 9 G and incidental (not intentional) electric fields of less than 1.2 V/m. In the first task, the monkey's estimation of a 5-second interval in response to a green light was measured by an "interresponse time schedule." In the second task, the subjects had to lift the lever repeatedly for a fixed number of times, in the presence of a red light, in order to get a reward of food or water (a "fixed ratio schedule").

The interresponse time (IRT) data showed no field-related changes in the mean values. However, two of the animals showed statistically significant ($p < 0.001$) variations in the distributions of the responses when the 45 Hz field was on, compared with "field-off" sessions before and after the exposures. The "field-on" IRTs were changed in the direction of more responses at shorter times and fewer at long times, leaving the mean IRT unchanged. A change in the IRT distribution, although again with no change in the mean, was observed for 3 of the 4 monkeys observed in a second experiment at 45 Hz where the fixed ratio task was omitted. Again, the change was to more of the faster IRTs.

Similar experiments at 15 Hz found no change in the mean IRTs and contradictory changes in the distributions. One animal showed a significantly ($p < 0.05$) faster time and another a significantly ($p < 0.001$) slower IRT with no differences observed for the other two monkeys.

In another experiment, a magnetic field was activated only during a 1-second reward-enable period occurring between 5 and 6 seconds after the stimulus was presented. Monkeys sensitive to the field might have used it as a cue to achieve higher accuracy and more rewards. This was not demonstrated but once again, contradictory changes in the distribution of the IRTs were found to be statistically significant for some of the animals.

The animals' rate of activity during the sessions was monitored with no field-related changes.

In summary, these magnetic field behavioral experiments were entirely negative except for changes in the IRT distributions.

deLorge, 1972; 1973a; 1973b

This series of 3 experiments on 2 rhesus monkeys exposed to a combined electric and magnetic field found no consistent effects on their behavior. In Experiment 1, the ELF fields were a 10 G magnetic field and a 4 V/m (peak-to-peak voltage) electric field at 75 Hz. In Experiment 2, a 10 G magnetic field and electric fields of between 3 - 7.4 V/m (depending upon orientation) varying at 45 Hz and 10 Hz were used. In the third experiment, a magnetic field of 10 G and electric fields of between 2.6 and 29 V/m (depending upon orientation and frequency) at 10 Hz and 60 Hz were used. Throughout these experiments, an ambient 60 Hz field of several volts per meter was present during exposure and control sessions. The magnetic field was produced in a pair of Holmholtz coils while the electric field was produced between 2 electrodes made of copper mesh.

The animals' behavior was examined with three tests: fixed interval; reaction time; and match to sample. In the fixed interval task, the monkeys responded to a green light by lifting a lever so that the monkeys' "willingness" to work could be measured by the rate of responses. In the reaction time task, the animals were trained to respond to a red light by lifting a lever and a subsequent audible tone required releasing the lever within 3 seconds. In the last task, the animals were trained to press a disc with a symbol matching one presented to the animal. The tasks were presented sequentially in 15-minute segments throughout an experimental session lasting 3 hours. Nine sessions with a magnetic field only and 6 with combined electric and magnetic fields were made for comparison with control sessions preceding and following the field exposure session.

The data at 75 Hz disclosed no field-related effects.

The data at 45 Hz gave essentially negative results with 4 isolated instances of statistically significant findings which the author explains as either artifacts of equipment malfunction or complications due to learning behavior shown by the animals.

A repetition of the data at 45 Hz and data at 10 Hz using the same subject animals as above, but longer sessions (generally, 8 hours) showed no combined field effects at 45 Hz. The 10 Hz fields did produce changes in the animals' behavior in an inconsistent, and statistically not significant, manner. A statistically significant decrease in activity levels ($p < 0.0$ was found for both animals (10 Hz) and this was the only finding that occurred in both subjects.

The suggestion of 10 Hz effects prompted a 10 Hz exposure to combined electric and magnetic fields for 8-hour periods on 24 consecutive days (no weekend break). These data showed that for 1 animal there was a significant ($p < 0.002$) reduction in the reinforcement times (the interval between food availability due to correct responses and action to get the food). Otherwise neither animal showed field-related effects in the fixed interva task nor significantly different performance in the action time task.

Both animals responded faster to the matching of symbols task (MS) when the fields were on, to the extent that 1 animal had a significantly faster response time ($p < 0.05$) and the other animal made fewer errors. Some of the other findings of the earlier 10 Hz experiments (on activity) were not confirmed.

The author's summary of the experiments at 45 and 10 Hz is that: "Few consistent patterns of behavioral change occurred as a function of the ELF fields, either between or within subjects, and only in the case of 1 animal was a repeatable effect observe However, this effect, a reduction in MS curves when the 10 Hz fields were on, was not statistically significant in any case." The authors point out that other inconsistencies in the experiment were not unusual, considering how many factors they examine they offer the explanation that there is a likelihood of a few chance occurrences in an experiment measuring so many parameters However, one positive finding at 10 Hz, reduced activity levels, was "highly unlikely as a chance phenomenon." Their failure to discover the same reduced activity level in the second 10 Hz experiment may have been "because of the overall reduction in activity that occurred as a function of increased confinement. This is, activity was reduced to such a low level that the ELF fields or even a very strong stimulus would have no discernible effect...The activity change in the current study was probably the only behavior presently explored that could have been an effect of 10 Hz fields." Their final comments emphasize the lack of consistency and the need for further determination of specific frequencies at which effects are notable.

The last of the three experiments by deLorge used 10 Hz combined fields and, in separate experiments, 60 Hz fields at about 10 G (magnetic field) and 2.6 - 30 V/m (electric field), the exact values depending upon frequency, orientation, and the particular exposure apparatus. Two female rhesus monkeys (previously unexposed) performed the 3 tasks used previously. The results were entirely negative for changes in task performance, general activity level, or any physical health change disclosed in a laboratory analysis of the blood.

The summary concludes: "...the present study, in addition to previous work, provides cumulative evidence that ELF non-ionizing radiation as produced in our laboratory does not influence animal behavior in either a beneficial or detrimental manner."

7.5 EEG AND ECG CHANGES IN A 50 Hz ELECTRIC FIELD

Blanchi et al., 1973b

Significant changes in the electrocardiogram (ECG) and electroencephalogram (EEG) patterns of animals were seen during and after exposure to a 100 kV/m, 50 Hz electric field. Mice were exposed in plastic cages with metal electrodes at the top and bottom in a cycle of 9 hours with the field "on" and 3 hours "off" for a total field exposure of 1,000 hours. ECG recordings were made on the exposed and control mice after 500 and 1,000 hours of exposure. The 1,000 hour results (Table 7-4) indicate a significant increase in the P-R interval (20%) and similar increases in the R-wave duration and QRS complex duration. The change in pulse rate was not reported.

A 0.5 hour exposure of guinea pigs to the 100 kV/m, 50 Hz field demonstrated "marked sinus arrhythmia beginning 10 - 20 minutes after removal of the field and lasting about 10 minutes."

The EEG activity of the guinea pigs exposed to the field for 0.5 hour was observed to have the following changes: 1) appearance of slow-frequency high-amplitude waves which persist during the whole exposure to the field; 2) increase in the number of spindles during the exposure to the field; 3) short-term desynchronization, which occurs soon after the application of the field." In addition, a short period of desynchronization was often seen shortly after the field was turned off.

7.6 EEG CHANGES IN AN ELECTROSTATIC FIELD

Lott and McCain, 1973

Static electric fields of 10 kV/m and pulsating fields (640 Hz for 0.1 second) were used for short-term exposures on

TABLE 7-4. ELECTROCARDIOGRAM CHANGES IN MICE EXPOSED TO A 100 kV/m, 50 Hz, ELECTRIC FIELD *

	PR interval	R wave duration	QRS complex duration
exposed	45.24 ± 1.40 ms	6.95 ± 0.38 ms	9.17 ± 0.48
controls	38.02 ± 0.10	5.59 ± 0.24	7.71 ± 0.25
difference	+7.22	+1.36	+1.46
% change	+19%	+24%	+19%
"p"	<0.01	<0.01	<0.05

* Blanchi *et al.*, 1973b

rats to examine for possible changes in the EEG as an index of awareness to the fields. Although no effects were found in respiration or rectal temperature, the EEG results correlated with the presence of the electrostatic field, showing an increase in brain activity with a decrease in hypothalamic activity.

The pulsating electric field also increased brain activity but in a more erratic manner, which also returned to normal when the field was turned off. In the pulsed field, the hypothalamic activity increased and was marked by a very strong reaction as the field was turned on, which was sustained throughout the field exposure and returned to normal when the field was turned off.

Experiments on a dead animal indicate that their findings are not in any way due to pick-up of the fields through the leads.

The conclusion is that the rats are clearly aware of the electric field environment and that the electrosensitive region of the brain is in the posterior hypothalamic regions.

7.7 EEG CHANGES IN MONKEYS IN HIGH MAGNETIC FIELDS

Beischer and Knepton, 1966

EEGs recorded for 1 - 3 minutes during exposure of monkeys to magnetic fields of 20,000 - 91,250 G, centered at the head, showed distinct changes in the amplitude and frequency of brain activity. Frequency components between 24 and 50 Hz, usually not prominent, were a major feature of EEGs recorded in the field and their amplitudes were 50 - 400 μV in contrast to the baseline EEGs, where the major components were between 8 and 12 Hz at 25 - 50 μV. Recordings made in homogeneous, gradient and reversed polarity fields were indistinguishable. The possibility of artifact is discussed and largely dismissed. In consideration of this and other research, the authors were able to conclude that "strong magnetic fields have a significant effect on the central nervous system..." although certain inconsistencies arose in considering the research then available.

7.8 ELECTRIC FIELD EFFECTS ON NEURONS

The experiments above suggest that external electric fields have effects on the operation of the central nervous system, a massive collection of neurons. We now discuss two experiments which indicate effects on individual neurons.

Gengerelli and Holter, 1941

The sciatic nerve of a frog, still connected to the gastrocnemius muscle, was placed in an electric field of 55 kV/m at 60 Hz to determine that the nerve can be stimulated by an external field as evidenced by a twitching of the muscle.

Severing the nerve from the muscle prevented any response, indicating it was a direct effect on the nerve, not the muscle.

Secondly, they noted that if the preparation was placed in a saline solution, the "electrolyte surrounding the nerve acted an electrostatic shield and prevented the lines of force from disturbing the electrical equilibrium of the nerve." Next, they tried to induce convulsive twitching in rats in an electric field of 275 kV/m but were unsuccessful; "possibly the fluids of the body were acting as an effective electrostatic shield."

These results are in accord with the calculations by Schwan in Section 4.10, where it is shown that direct nerve stimulation in the whole animal requires still larger external fields.

Terzuolo and Bullock, 1956

The effects of an electric field on an exposed nerve cell, the stretch receptor of a crayfish, were studied using a nerve and muscle preparation in a mineral oil bath. By stretching the muscle fibers, the receptor fired rhythmically at a rate of between 5 and 15 impulses per second. By passing a current through the preparation, the rate of firing could be changed by up to 72%, although the current was so small that an adjacent nerve, not responsive to the muscle tension, was not induced to discharge. Thus, the behavior of a neuron can be changed by an electric field very much smaller than that necessary to fire a quiescent neuron. The authors estimate that the voltage gradient across the cell body (soma) is of the order of 1 V/m (an upper limit). Contrasting this electric field gradient with the gradient across the cell membrane, which is some 10^7 times greater, it is difficult to understand how such a small perturbation could affect the nerve's function. The authors are not able to pose a fully acceptable explanation of the results, but indicate that the cell must be sensitive to differences in the membrane potential from point to point, rather than the potential across the membrane.

7.9 MITOTIC CYCLE IN SLIME MOLD

Marron et al., 1975; Goodman et al., 1975

Slime mold (*Physarum polycephalum*) can be grown in synchronous cultures with great regularity of the mitotic cycle at a rate of approximately 3 divisions per day. Cultures in shake flasks and stationary cultures on filter paper were exposed continuously to electric and magnetic fields of 60 or 75 Hz at levels of 2 G and 0.7 V/m generated by a pair of rectangular coils (for the magnetic field) and by a pair of stainless steel electrodes in contact with the medium (producing a current corresponding to the electric field). Possible electrolytic effects were not found in dummy flasks or culture dishes containing only the growth medium.

The experiment measured the time for a second mitotic division after the addition of growth medium, normally about 14 hours. It was confirmed that this interval was an accurate measure of cycle times by measurements from the second to the third mitosis made in auxiliary experiments.

The Sanguine-level electric and magnetic fields caused a significant ($p < 0.05$) lengthening of the reproductive cycle in a culture of slime mold observed over two years. There was a significant difference between the 60 Hz and 75 Hz fields in that the 60 Hz fields produced mitotic delay in 80 - 100 days, while it took 100 - 120 days for the same effect to occur in the mold exposed to 75 Hz fields. It was determined that the effects were reversible over a period of 30 - 60 days after removing the fields.

No effects were obtained using 75 Hz fields at reduced levels of 0.4 G and 0.15 V/m.

7.10 FERTILITY AND PATHOLOGY IN MALE MICE

Knickerbocker et al., 1967

Twenty-two male mice were exposed to a 160 kV/m, 60 Hz electric field for a 10.5 month period to test for possible effects on the reproductive mechanism by a comparison of their progeny with the progeny of 22 control animals. The mice (housed 1 per cage or, in 4 cases, 2 per subdivided cage) were exposed in a metal covered acrylic cage, placed between mesh electrodes (hardware cloth) located above and below the 10 cages. The electrode was energized to 24 kV, necessitating a smooth border-wire to eliminate corona at the wire ends. Exposure was for 6.5 hours per day, 5 days per week, for 10.5 months, with access to virgin breeding females occurring during the non-exposure period. The large field strength resulted in an audible corona discharge when the animals stood up on their hind legs and "discouraged" the mice from using the water bottle's glass nipple. For the reason, the exposure period was interrupted by a 40-minute "water break" during which the field was off.

Each male bred two females per month and 8 of the litters produced each month were studied after being trimmed to 4 pups each. The weight of each mouse was recorded weekly from birth to 90 days of age; after weaning at 25 days, the male and female progeny were separated. The 22 control males were identically treated except for the electric field and consequent water deprivation.

The authors note some behavioral changes in the exposed animals who, while in the field, confined themselves to the cage bottom, were disturbed by the field when they would stand on their hind legs, and seemed to sleep more than the controls.

The exposed and control mice did not differ in weight according to measurements made in mid-study and at the end. The weekly weighings of the progeny of the experimentals and controls did not show a difference among the females, but there was a significant ($p < 0.05$) 4% reduction in the average weight of the male progeny, observed throughout most of the 90-day period.

The number of litters per month was unaffected by the field except for a sharp decrease in the second month, which was belived to be an artifact that was not reproduced upon repetition of the first 2 months of the study.

Histopathological studies of half the exposed males and controls were made at the end of the study and showed no relevan pathology.

The authors suggest that the weight difference among the male progeny may be related to housing conditions which placed the progeny of the exposed males opposite a window and the control's progeny opposite a wall. Why only the males were affected is not explained.

7.11 ACTIVITY IN MICE

Moos, 1964

The effects of an alternating electric field (60 Hz) on the activity of caged mice was investigated at a field strength of "about 800 - 1200 V/m," using electrodes at the top and botto of a plastic cage. In 5 replications of the experiment involving exposure periods of 1 month (1 run), 5 days (3 runs), and 1 day (1 run), alternating with control periods of equal length, significant differences in the mouse activity rates were discovered. The results (significant at levels between 0.5 and 5%) indicated that the experimentals were more active during periods of darkness with the electric field applied, and (with one exception) less active than the self-controls during periods of light. (Mice are nocturnal, and much more active during darkness.)

A second experiment to study the relation between electric fields and mortality in rats pre-stressed with X-radiation of 900 R was inconclusive. Although there was greater mortality in the exposed mice, beginning on the fifth day after irradiation, only occasionally was the difference of statistical significance (5% confidence level).

7.12 EFFECTS ON MICE

Belyaeva and Stroikova, 1975

Laboratory studies on mice disclosed "metabolic disorders, lowered resistance to infection, paralysis, necrosis, and late deaths from secondary causes which did not occur in control animals." An experiment on rabbits showed effects on the "compensatory mechanism of thermoregulation" and an elevation in blood pressure. Unfortunately, this report, known only from the translator's abstract, does not specify the electric field exposure condition.

7.13 PSYCHOLOGICAL BEHAVIOR

Persinger and Pear, 1972; Persinger et al., 1973

Persinger and Pear report an experiment at 0.5 Hz, in which rats are exposed prenatally to the magnetic field obtained by rotating a permanent horesehoe magnet ("rotating magnetic field"). They found that at 70 days of age, the rats which had been exposed *in utero* demonstrated significantly different behavior in an experiment which required them to suppress a lever-pressing response previously encouraged. The rats exposed to the ELF fields responded more quickly to the conditioned suppression of their trained behavior ($p < 0.01$) on the third and fourth days after suppression was begun, but this relatively greater suppression disappeared by the next few days. The magnetic field intensities varied from 0.5 - 30 G, with the suggestion of some field intensity-related effects, but no quantitative demonstration.

The rotating magnetic field is used to simulate low level atmospheric magnetic field variations due to geophysical and meteorological changes which the author believes may bear on development and behavior.

In a literature review (Persinger *et al*., 1973), rotating magnetic field experiments were reported to have affected ambulatory behavior, defecation, and to have caused reduced avoidance of an electric shock. Another experiment using rats exposed to a rotating magnetic field during the first three days after birth indicates transient effects in behavior which the author associates with changes in the cerebellum. He notes that the cerebellum undergoes maximal development during the days used for the exposure and speculates that a change in oxygen uptake due to the ELF fields may be associated with the effects.

Persinger, 1974

ELF psychological experiments and phenomena such as correlations between behavioral experiments and the lunar distance are reported. The author discusses the interaction of thyroid activity, the autonomic system, oxygen consumption, and lipid metabolisms in a manner suggesting correlation between results previously found and reported in the literature. Persinger also suggests that the ELF magnetic fields have a direct influence on the nervous system, probably at the vagus nerve, and demonstrates that a 0.5 Hz magnetic field induces currents of the order estimated to be biologically effective in the experiments by Becker. (See Section 7.25.) Finally, in a summary article, Persinger discusses mechanisms by which ELF fields may affect biochemistry and biobehavior. He discusses direct interaction with the nervous system, interactions through phase changes in liquid crystal states of lipids in the cell membrane, and perturbations in the polyanionic sugar groups related both to membrane phenomena and immune response effects.

7.14 ELECTRIC AND MAGNETIC FIELD EFFECTS IN ANIMAL MIGRATION, NAVIGATION, AND COMMUNICATION

The behavior of migratory birds and fish has long been a mystery, which is just now being unraveled. The earth's static magnetic field is one of several cues which fish and birds use during migration. In addition, some fish have been shown to use electric currents for communication and to sense the presence of prey, etc. The remarkable sensitivity these animals show and the remarkable organs which have been adapted to achieve these sensitivities are of interest, at least insofar as they indicate techniques of field reception and levels of sensitivity

Rommel and McCleave, 1972; 1973a; 1973b

American eels and Atlantic salmon are sensitive to very low level electric fields and it is hypothesized that this sensitivi enables them to use the electric fields developed in ocean currents to guide their long-distance migrations. Sensitivity to low level electric fields was demonstrated in a behavioral experiment using DC electric fields of varying strengths as a conditioned stimulus and observing a change in the heartbeat rat which occurred in response. After a 3-heartbeat-long exposure, to a DC external field, an AC electric shock was given as the unconditioned stimulus. The fish were exposed in tanks, using flat plate electrodes to establish the fields and also to provid the unconditioned stimulus shock. The EKGs were monitored with surgically implanted electrodes.

By observing the heart rate for 3 beats before and 3 after presentation of the DC field, it was found that both salmon and eels are sensitive to electric fields perpendicular to their bodies when the electric fields and water conductivity (varying with the water's salinity) produce sufficient electric current densities. For the American eels, results were highly significant for current densities of 16.7 $\mu A/m^2$ or greater in fresh water (resistivity: 40 ohm-meter) corresponding to electric fiel of 67 $\mu V/m$ or more and a power density of 1.1 nW/m^3. Even at the largest current densities and greatest electric fields in fresh water (1.67 mA/m^2 and 67 mV/m) the fish were sensitive only to a perpendicular electric field and not to the parallel field. By increasing the salinity of the water at fixed current density fewer eels could be trained, demonstrating the presence of an electric field threshold. Similarly, reducing the current density indicated a current density threshold so that, for a giv electric field, the fish show greater sensitivity as the salinit of the water (and conductivity) increases, consistent with the idea that this sensitivity is useful for fish migrating in sea water. Threshold current densities are of the order of several microamperes per square meter (2 - 20 $\mu A/m^2$) and threshold electric fields are in the range of several tens of microvolts per meter (5 - 50 $\mu V/m$), depending upon the water resistivity. (See Table 7-5.) The data suggest that the response threshold is more accurately characterized by the power density.

These sensitivities are apparently adequate for the fish to sense oceanic currents through the following mechanism. From Faraday's Law the vector product of the water's velocity with the vertical component of the earth's static magnetic field results in an electric field perpendicular to the direction in which the water mass is flowing. By calculations based upon known velocities and resistivities for the Gulf Stream off the southeastern coast of the U.S., horizontal electric current densities up to 170 $\mu A/m^2$ and horizontal electric fields up to 46 $\mu V/m$ may occur, indicating that the fish are adequately sensitive to these horizontal fields to use them as guides when swimming along with the current. The authors also note that in the side-to-side motion of swimming, rapid changes occur in the electric field relative to the fishes' body, which suggest the fish is sensitive to an AC signal. Other research referred to by Rommel and McCleave, 1973, has demonstrated phase sensitivity in other fish.

Bullock, 1973; Hopkins, 1974

Several species of marine life, both fresh and seawater varieties, have highly sensitive electroreceptors and, in many cases, they also possess electric organs which transmit electric signals of either a wavelike or pulsatile nature. These fish are found in both the sea and fresh water and respond to electric fields from DC to several kilohertz. Fish sensitivity to electric signals may be used to locate other fish for the purposes of gaining food, responding to threat, or mating. Fish have also been demonstrated to be sensitive to atmospheric electricity, for example due to thunderstorms, and might be able to sense manmade fields such as result from current return paths in the ground.

Seawater fish have developed receptor organs utilizing long tubules (up to 1/3 the body length) which magnify the potential difference between the seawater at the pore opening and the body potential within the fish. In freshwater fish, the relatively high resistivity of the water makes this amplification scheme unnecessary and the receptors are very short in length. The electroreceptors are further classified as "ampullary," responding to low frequencies (below 50 Hz) or "tuberous," responsive above 50 Hz.

In fish equipped with transmitter capabilities, the electric organ arises from either muscle or nerve tissue organized into

TABLE 7-5. EEL AND SALMON HEART-RATES CONDITIONED BY AN ELECTRIC FIELD*

Electric Field ($\mu V/m$)	Water Resistivity (ohm-m)	Current Density ($\mu A/m^2$)	Power Density (pW/m^3)	No. of fish showing significa[nt] conditioning/total tested ⊥	∥
a) American Eels					
6,700	40	167	$1{\cdot}1 \times 10^6$	6/7	-
670	40	16.7	$1{\cdot}1 \times 10^4$	6/8	0/4
67	40	1.67	110	2/8	-
67	4	16.7	$1{\cdot}1 \times 10^3$	8/8	0/4
6.7	40	0.167	1·1	0/8	-
6.7	4	1.67	11	0/8	-
6.7	0.4	16.7	110	3/8	-
b) Atlantic Salmon					
600	40	15	9×10^3	6/10	0/3
60	40	1.5	90	5/9	0/3
60	4	15	900	5/9	0/3
6	40	0.15	0.9	3/9	-
6	4	1.5	9	4/9	-
6	0.4	15	90	5/9	-

The cardiac deceleration was studied with an electric field applied either parallel (∥) or perpendicular (⊥) to the body axis of the fish. Significance was determined by a "p-value" of 0.05 or less on a one-tailed t-test.

* Rommel and McCleave, 1973a

long bundles of cells in the tail region. During the discharge of the organ, potentials generated in individual cells are summed to produce a relatively large voltage (less than 1 V), resulting in a dipolar electric field. These fields can be detected using electrodes and an audio amplifier, oscilloscope or other transducer.

In sharks and rays, the structures known as the "ampullae of Lorenzini" were long a mystery until, in 1939, they were found to be highly temperature-sensitive. It took over 20 years more to discover that, in addition, the ampullae of Lorenzini enabled the shark to detect prey buried in the sand by detection of a feeble, steady electric field which emanates from the buried prey. It was shown that the shark detects and aims an attack at a flat-fish even if covered with an agar plate which prevents mechanical, visual, or chemical cues, but allows the electrical cue. The shark will even attack buried electrodes which create a mock flat-fish current.

Electroreception is sensitive enough to detect animate and inanimate objects over a range of several meters corresponding to a threshold sensitivity of as low as 1 μV/m. Individual fish show idiosyncratic transmission frequencies and different species operate in different frequency ranges. The peripheral and central nervous systems are now being studied in order to understand the organization of the electric communication facilities.

The information content of various wave forms and various frequencies or combinations of frequencies is also under study, with results such as the remarkable discovery that one fish (Sternachus) maintains the output from its electric organ to an accuracy of 1 part in 10^4 over several thousand cycles.

In summary, many species of fish have developed the ability to receive electrical signals useful for detecting prey or predators, possibly useful in sensing changes in water salinity, ocean or river currents, and geomagnetic or geoelectric events. In addition, many other species are both electroactive and electroreceptive, enabling communication for the purposes of remote identification of animate and inanimate objects, which is useful in seeking food and mates or in avoiding enemies.

Southern, 1973; 1975

In previous works, Southern observed that ring-billed gull chicks showed an unlearned orientation preference at a compass bearing toward the southeast, the normal migratory heading. This preference apparently depends (in part) on a cue from the static earth's magnetic field. An 8 ft. diameter cage confining a single 3 - 9 day old bird was placed under the influence of electric and magnetic fields associated with Project Sanguine test antennae. For most of the experiments, in the influence of a buried antenna, operating at 45 or 76 Hz, the magnetic field in the cage was about 0.5 G and the electric field 0.2 - 0.3 V/m. Tests were made using an unmodulated ELF carrier signal and signals modulated at 4 Hz.

Individual chicks were placed in the center of the cage under overcast as well as clear skies to control for possible solar cues. Their heading at the end of 2 minutes or when they reached the outer wall of the test chamber was observed through one of 2 portals. First, it was established that the control groups -- antenna fields off -- had highly significant orientations at a mean heading of about 165°, regardless of solar cues. Birds exposed to the 76 Hz unmodulated ELF fields under a clear sky no longer had a significant mean heading, indicating disorientation, but birds exposed under an overcast sky still maintained significant mean bearings. All other test conditions (45 Hz and 76 Hz modulated at 4 Hz, or 45 Hz unmodulated and either overcast or clear skies) indicated disorientation effects Thus, the only category for which disorientation was not observed was at 76 Hz for unmodulated waves under overcast conditions, where the birds maintained a mean heading that was significant at the 0.0005 level, comparable to the controls. The author cautions that the orientation which persisted in the presence of the 76 Hz, unmodulated ELF field on overcast days may be suspect because few trials, involving only 2 groups of chicks, were possible under these conditions.

The author's conclusion is that: "The results strongly suggest that the magnetic, and possibly the electric, fields... cause a significant reduction in the ability of ring-billed gull chicks to select a mean heading predicted for birds from this population. The result, in some instances, was complete disorientation (randomness), whereas in other cases a significant mean angle existed in the data but it was skewed significantly from the predicted mean and thus seemed to lack biological significance."

The significance of these findings for birds flying at altitudes of 1,000 ft. or more above the antenna system, is not demonstrated, inasmuch as the magnetic field at those altitudes would be many times weaker.

Walcott, 1974

This article reviews the various mechanisms homing pigeons use in returning to their lofts from remarkably large distances (over tens or hundreds of miles) by use of innate abilities which are perfected by training the birds. Apparently, pigeons use several navigational techniques including recognition of landmarks, the sun's position as a compass indicator, sensing of odors, and magnetic fields.

The evidence for the importance of magnetic fields as one component of homing ability is demonstrated in a number of experiments. If small magnets are attached to the bird's wings, there is a clear effect on the bird's navigation in overcast skies and, for longer distances only, in sunny skies too. Using a battery-powered Helmholtz coil to produce a magnetic field of several tenths of a gauss near the bird's head, it was demonstrated that the birds could sense this magnetic field. If the bird is deceived by an artificial field to believe that the geomagnetic field is reversed in polarity, the bird heads off in the opposite direction. There seems to be adequate evidence for the role of a "magnetic compass" in homing pigeon behavior, but the general question of how birds do find their way home is not yet answered.

Blakemore, 1975

Certain bacteria found in marine sediments move in response to weak DC magnetic fields of the size of the geomagnetic field (0.5 G). Their response to a magnetic field may be related to the presence of chains of iron-rich, crystal-like particles found within membranes in the cytoplasm. The magnetotactic response depends upon the bacterium's ability to move (by flagellar motion) and is not a result of a direct magnetic force on the cell.

Graue, 1974

The possible effects of Project Sanguine electric and magnetic fields on the homing pigeon were investigated at the test transmitter site on 8 dates, using 161 birds. The first part of the experiment measured the time it took the birds to return to a loft located at the transmitter site and the second part measured the compass bearings and time it took to vanish from sight for birds starting at locations near the transmitter. The magnetic fields were 5.6×10^{-3} G and the electric field less than 0.1 V/m. Birds were released when the transmitter was off. The results did not demonstrate clear ELF field effects, but the author found some hints of changes that require further investigation. He concludes that if there is any effect at all, it is "only a matter of a slight alteration of headings while the birds are close to the antenna." He also seems to have found that small geomagnetic field fluctuation had a more serious effect on orientation in this experiment than did the Sanguine-produced fields.

7.15 ELECTRIC FIELD EFFECTS ON MARINE ANIMALS

Straub et al., 1972

The effects of ELF electric fields on a number of marine animals were examined in the laboratory using stainless steel plate electrodes in a plastic tank to produce current densities in the range of 10 - 650 A/m^2. The studies were conducted at frequencies between 10 Hz and 20,000 Hz. Each animal was observed for its response, e.g., withdrawal of tentacles, contractions of certain muscles, a bioluminescent flash, or rapid swimming. The animals investigated included the sea anemone, mnemiopsis, bay scallops, mud snails, crabs, shrimp, sea squirts, and fish. The most sensitive response was obtained from the green shrimp (Panaeus setiferus) which was sensitive at 75 Hz to current densities below 10 A/m^2. At low frequencies fish were stunned and the young ones died at current densities of 40 - 80 A/m^2, while larger fish recovered.

Despite the wide variations in the sorts of marine life tested, including large differences in the skeleton which affect the current density penetrating the organism, the range of threshold responses at 10 Hz was small, within one order of magnitude for all the organisms. By 100 Hz this range had increased appreciably to almost 2 orders of magnitude. Because of the similarity of the response curves found in all these different animals, and in consideration of other literature they cite, it is suggested that AC currents have their effect at the cell membrane.

7.16 EFFECTS ON CONDITIONED BEHAVIOR IN PIGEONS AND RATS

Marr et al., 1973

ELF fields at 45, 60, and 75 Hz at magnetic field strengths between 0.1 and 2 G and electric fields between 0.06 and 100 V/m were used in an experiment which found that operantly conditioned behavior of pigeons was not affected. Next, the possibility that pigeons or rats may be able to detect ELF fields was explored by using the fields as either discriminative or conditioned stimuli, again with negative results. Finally, using pigeons, it was determined that the birds expressed no preference between ELF field conditions or the control (no field) conditions.

The animals were exposed in cages surrounded by Helmholtz coils to provide the magnetic field and a pair of parallel metal plates produced a uniform electric field. The electric field fringing effects were reduced by using metal rods to enclose the field region, including use of the rods in the floor. Details of the caging and plots of the measured fields are presented.

7.17 EFFECTS ON GROWTH AND DEVELOPMENT

Krueger et al., 1972

This report examined the growth of chicks exposed to ELF electric or magnetic fields from birth until 28 days of age. Estimated electric fields of 3.4 kV/m (60 Hz) and 3.6 kV/m (45 Hz) or a magnetic field of 1.2 G (60 Hz) and 1.4 G (45 Hz) were used to expose chicks housed in metal cages, thus producing an unmeasurable, non-uniform field distribution. The electric fields were generated by applying 800 V between an aluminum cover plate and the cage. The results showed that chicks exposed to electric fields at either frequency had slightly lower rates of growth which were not statistically significant ($p > 0.05$). Statistically significant results were found for the 45 Hz magnetic field exposure ($p < 0.01$).

This experiment detected no differences in "livability," activity, or behavior.

Durfee et al., 1975a

Embryonic and newly-hatched chicks were exposed to ELF magnetic or electric fields at 45 , 60 and 75 Hz (magnetic intensities of 1, 5, 8, or 30 G and electric fields of 1, 10, or 3600 V/m) in an investigation of the growth and development of the embryo, growth and development of the newly-hatched chick, and finally, the growth of the young bird. Electric fields were produced between parallel plates and Helmholtz coils provided a uniform magnetic field for the incubators or cages used in the ELF exposure. The separate electric and magnetic field exposures occurred throughout the incubation period and until the chicks were 4 weeks of age.

A statistical analysis of the results showed that there were no significant ELF effects on the fertility, hatchability, or survivability of the exposed chick embryos compared with the control eggs.

Magnetic fields of 5 G at 60 or 75 Hz affected the respiration rate of 12 or 13-day-old embryos, causing a significantly decreased CO_2 production for embryos exposed continuously. The authors caution against "undue confidence in these results." Further experimentation resulted in another experiment at 60 Hz which did not confirm the effects on CO_2 production (Durfee *et al.*, 1975b).

A behavioral experiment which measured the young chick's inclination to peck disclosed no ELF-related differences.

Body weights 24 hours after hatching for chicks incubated in the fields were no different in the exposed and control populations. Continuing measurements of body weights at intervals of 1 or 2 weeks for a period of 10 weeks disclosed no differences due to the field, in contradiction with other research in the literature. The authors comment that the contradicting experiments differ in caging and exposure conditions because, in the report of positive effects on weight, the chicks were not exposed until after hatching.

Smith, 1975

Adult chickens, derived from an experiment (Durfee *et al.*, 1975a) in which they were exposed to a 45 Hz, 8 G magnetic field throughout growth and development, were examined for possible effects on social behavior as expressed in pecking order. In 2 of 3 trials involving 5 exposed and 5 control birds each, statistical analysis of the viewer's observation of pecking and threatening behavior disclosed that the ELF-exposed birds formed a more dominant population than the the control birds. This possible ELF effect would need further investigation to confirm its existence.

Yates et al., 1975

An *in vitro* experiment on chicken embryo fibroblasts and embryonic kidney cells exposed to ELF fields disclosed both growth inhibition and acceleration depending upon field modality. Fibroblast cells from 10-day-old chicken embryos were cultured so that the total number of viable cells could be observed at various times between 19 and 70 hours after the seeding of the culture. Cells derived from the culture medium by centrifugation and resuspension were counted in a hemocytometer as well as by a technique using tritiated uridine.

The embryonic fibroblast cells exposed to 60 Hz magnetic fields at 1, 5, or 8 G showed inhibited growth at a significance level of $p < 0.01$. Using embryonic kidney cells (from 21-day-old chicks), similar results were found, again significant at the $p < 0.01$ level. The radioactive labeling experiment showed similar results indicating inhibition of growth in the magnetic field ($p < 0.05$) for both the fibroblast and kidney-derived cells.

Using electric fields of 10 V/m at 60 Hz, growth was again shown to be significantly inhibited for both types of cells.

Using 75 Hz electric fields at both 10 and 1 V/m, conflicting results were found in that the 1 V/m-exposed cells showed inhibited growth ($p < 0.05$) while the cells exposed at 10 V/m indicated enhanced growth compared with the controls ($p < 0.05$). These results were confirmed by the radioactive labeling experiments, except that the enhancement effect was not statistically significant in this case.

The authors are unable to explain the observed effects.

Durfee et al., 1975b

In view of the above experiments, which indicate *in vitro* magnetic and electric field effects or *in vivo* magnetic fields effects at 5 G, embryo respiration was measured by the rate of CO_2 production 6, 13, and 19 days after incubation began. The incubating embryos were exposed to an ELF magnetic field of 5 G at 60 Hz continuously except for 1 hour during the measurement period. The results from this experiment are entirely negative, disclosing no significant differences between ELF-exposed chick embryos and the controls.

7.18 ELECTRICAL ACTIVITY OF EARLY EMBRYOLOGICAL STAGES

Jaffe, 1966; Robinson and Jaffe, 1975; Weisenseel et al., 1975; Novak and Bentrup, 1973

The patterns of growth in the embryonic stages following shortly after fertilization of the egg cell provide evidence that electric factors are important in early development. Polarization of the cell mass, perhaps due to self-generated potentials following changes in cell membrane conformation, may be responsible for the developmental axis which soon marks the dividing embryo. Jaffe uses an experimental system based on an alga (*Fucus furcatus*) because of their normally pattern-free development. Normally, about 12 hours after fertilization, the zygote starts to elongate by bulging at one pole; then becoming pear-shaped, it divides into a rhizoid portion which holds on to the substrate and another cell called the thallus. The rhizoid from then on locates the pole of the organism's future development. It has been demonstrated that light (and a number of other agents) can influence the location of this pole at any time *before* the rhizoid develops.

Using this light sensitivity to orient egg calls, Jaffe was able to demonstrate the development of a negative electrical pole at the rhizoid end of the egg. Eighteen-and-a-half hours after fertilization (about 6 hours after the start of the signal marking formation of the rhizoid) there is a voltage of about 2.7×10^{-7} V per egg, corresponding to a calculated current through the egg of 6.4×10^{-11} A. The variations from one sample to the next are small. In his experiment (1966) Jaffe goes on to demonstrate by means of rapid perfusion and an analysis of the electrical properties of the experimental set-up that the potentials arise from the membrane and not from an extracellular ionic gradient.

Although it was difficult to continue embryonic development past the 4-cell stage, from the data up to 4 cells it appears that the embryonic potential increases in magnitude throughout the period from fertilization to the 4-cell stage. Thereafter, due to the experimental situation, no further division occurred and the electric potential started to decline in magnitude.

The author points out that although the potential outside the cell is relatively negative in the region of elongation (the rhizoid end), the usual membrane potential is positive at the outside of the cell with respect to the interior, thus implying that the membrane in the region of elongation is depolarized relative to the membrane elsewhere. He estimates that the current density is 60 mA/m^2, which corresponds to an internal electric field of 0.13 V/m -- "a marked underestimate." This electric field may then affect the ion gradients in that region of the cell membrane with consequent effects on the cell itself. Thus, there is the suggestion of a feedback loop in which the cell generates the potential and the potential alters the cell in its turn.

Robinson and Jaffe (1975) have demonstrated that the current is due to a calcium ion flux, in which the cations enter the rhizoid end in greater number than at the thallus end, resulting in an electro-negative region exterior to the rhizoid. The calcium current measured in this experiment was 2×10^{-12} A per egg, 6 hour after fertilization, decreasing to 0.4×10^{-12} A 10 hours after fertilization (just about the time the rhizoid develops). This implies that at 6 hours the calcium flux represents most of the current, but by 10 hours the calcium flux is only a small fraction of the total current. As suggested in other experiments, this indicates that sodium, chlorine and calcium ions may all be involved in the polarization process.

Another experiment from Jaffe's laboratory on the early growth of pollen grains in a low conductivity medium (Weisenseel et al., 1975) measures the regions of the current source and sink. They find maximum current densities of 0.6 A/m^2 producing an internal electric field of 1 V/m within the pollen tube's cytoplasm.

Novak and Bentrup use AC and DC fields to cause the alga eggs to develop with a particular pole orientation, rhizoid toward the cathode. The peak response for an AC signal was found at a pulse rate of 8 Hz, using pulses about 10 meters in length. The authors interpret this result to mean that electric field-induced cell polarity (unlike the light-induced polarity in the above experiment) has its effect by shifting the membrane potential rather than by causing a current flow through the cell. In this experiment a dose-response curve is obtained which is "strikingly similar to other morphogenetically active signals."

Novak and Bentrup suggest that "electrostriction of the plasmalemma structural proteins" may be the mechanism by which electric fields are sensed in the membrane.

* * *

The following sections draw heavily from pilot studies which were conducted for Project Sanguine to determine the need for further research. A wide spectrum of life forms from bacteria to mammals were exposed to electric and magnetic fields of 10 or 20 V/m and 1 or 2 G at 45 or 75 Hz.

7.19 MULTIGENERATIONAL STUDY IN RATS

Coate et al., 1970

Albino female rats were exposed to combined electric and magnetic fields throughout their life cycle. The experiment included production and rearing of a second generation, which was also continuously exposed up to the age when their offspring were weaned, and kept offspring data on the number of live and still births, mortality, and body weights at 3 intervals in time. Histologic studies were undertaken, showing no significant pathology. Slightly lower fertility for the exposed group ($p < 0.1$) and a significantly higher lactation index for the exposed population ($p < 0.001$) were indicated.*

The authors note that the overall lactation index was exceptionally low, probably due to the peculiarities of the housing conditions. The difference in lactation indices occurs only for the second generation of rats raised in the fields. The data are of low reliability in view of the wide fluctuations in the first and second generation experiments under all the exposure conditions.

7.20 BEHAVIOR AND PHYSIOLOGY OF DOGS

Coate et al., 1970

The behavior and physiological function of dogs were studied in an experiment which exposed the animals 16 hours per day, 5 days per week for 3 weeks. The same ELF field conditions as above were augmented by a 0.5 mA current (AC) carried through their bodies between leg-electrodes. The data were intended to be similar to those acquired in the study on the health of utility linemen (Kouwenhoven et al., 1967). Twelve of the 14 dogs (equal numbers of males and females) were operated on to allow direct arterial blood pressure measurements and EEGs. Two animals (male and female) were studied at each of the exposure conditions. The results, which included extensive laboratory exams and autopsies, were negative with the exception of "possible effects" on rectal temperature (elevated in 2 males exposed at 20 V/m, 2 G at 75 and 45 Hz, respectively) and for upward shifts in the blood pressure affecting 3 of the 4 dogs exposed at 45 Hz and all 4 exposed at 75 Hz. These changes were noted on one or more occasions during exposure or post-exposure periods.

The authors doubt that the elevated temperature measurements are a real field effect, but place more emphasis on the hypertensive effect occasionally seen in both the systolic and diastolic blood pressures. Again, it should be noted that in addition to the electric and magnetic fields in this study, a current of 0.5 mA was passed through the animals.

7.21 MUTAGENESIS IN *DROSOPHILA*

Coate et al., 1970

The possible mutagenic properties of ELF fields were explored in a 48-hour exposure of male *Drosophila* to fields of 20 V/m and 2 G at either 45 or 75 Hz. In this experiment, an alteration in the exposed male's chromosome shows up in the second generation by an absence of "wild type" flies, whereas without mutations, the wild type flies are found in approximately equal proportion to the other type ("Muller-5").

---

* The lactation index is a fraction of pups alive at 21 days compared with those alive at 24 hours, expressed as a percentage.

The results indicated mutation in that no control cultures were without the wild types, but 8 of 150 cultures exposed at 45 Hz and 7 of 150 exposed at 75 Hz lacked the wild type male fruit flies. This result is significant at $p < 0.001$, so that the authors conclude: "Results strongly indicate a mutagenic action of the EM fields." The authors discuss the possibility of artifact, with the conclusion that its probability was very low, and their final statement is: "Due to the momentous biological implications of a strong dominant mutagenic effect from weak electromagnetic field exposure, acceptance of the present results as demonstrating such an effect would require replication in much larger samples. Also a dose-related effect should be demonstrated before concluding that ELF electromagnetic fields can be mutagenic."

Mittler, 1972

In response to the surprising report of mutagenesis in *Drosophila* due to ELF fields (Coate *et al.*, 1970, and commonly referred to as "the Hazleton study), a similar experiment was performed at 75, 45, and 60 Hz, using electric fields of 10 V/m and a magnetic field of 1 G. After a 5-day exposure to the ELF fields, males were mated to 3 successive female groups in successive 2-day intervals, thus yielding broods from sperm material which was either mature or immature at the time of exposure. The genetic make-up of the flies in this experiment were explained in detail and consideration was given to the possibility that an X-chromosome mutation might occur, resulting in the failure of the wild type males to reproduce. The results of this experiment were entirely negative. The author suggests that the effects reported by the Hazleton study may be explained by failure to use virgin female flies, although Coate *et al.* described their technique for insuring virginity as "rigorous." Mittler goes on to comment that in other experiments in his laboratory, magnetic fields of 11,000 G were found to be non-mutagenic too.

7.22 MUTAGENESIS IN *E. COLI*

Coate et al., 1970

Mutagenesis in *E. coli* was investigated in a culture placed in a plastic vessel and subjected for 24 hours to a 45 or 75 Hz magnetic field of 1 or 2 G and an electric field of 10 or 20 V/m produced by a current of 0.5 mA. Mutations were indicated by the development of a strain unable to grow in the medium provided and were detected by a selective killing of the growing bacteria, leaving any non-growing mutants for analysis. Under all test conditions, 264 colonies were found to have mutant bacteria, distributed between the controls and exposed fields in a manner that indicates the ELF fields are not mutagenic. The validity of this technique was shown by using a known mutagen, ultraviolet radiation, which did develop a significant number of altered bacteria.

7.23 PERCEPTION IN SEVERAL SPECIES

Coate et al., 1970

Possible detection of ELF fields by fish, ducklings, turtles, rats, and fruit flies was studied in pilot experiments. Using the above exposure conditions in air or tap water, detection was to be shown by aversion to the electric or magnetic field regions. After a period of acclimation, the fields were turned on for an hour, at which point the animal's position was determined. Depending on its location, the fields were energized in one or the other region and, after another hour, the animal's position was again noted.

Aversion behavior was observed only for turtles at 20 V/m. At 20 V/m, fish demonstrated perception by their changed swimming motions but not aversion, perhaps because they are able to reduce the potential from head to tail by turning their bodies to be perpendicular to the field. The authors' conclusion is that an electrical field of 10 V/m maintained in either the earth or water was not "noxious" enough to cause aversion, although in water such a field may be close to the threshold for effects.

The behavior of rats, as demonstrated in learning how to avoid a shock, was used to examine the offspring of rats which had lived in the field for at least 80 days prior to conception. The offspring themselves were raised in the field until about 30 days of age (at weaning). This experimental protocol tests for possible nervous system effects of ELF fields and was entirely negative in its findings for exposures at 45 and 75 Hz, 1 and 2 G, and at 20 V/m.

Milburn, 1971

This report, which is a part of a thesis dissertation, includes a comprehensive literature survey from which we draw the following citations. An experiment was performed showing that hamsters avoid the region of an electric field placed directly above their nest in the frequency range of a "few hertz to about 10 kilohertz" and at strengths of 100 - 900 V/m (Schua, 1953; 1954). A vertical electric field of 1.75 - 5 kV/m affected the rate of locomotion in a snail (Webb *et al*., 1959) and a 200 V/m horizontal electrostatic (DC) field affected the paths of snails causing a left-right asymetry in their movement (Webb *et al*., 1961). The planarian *Dugesia* perceived a change of 200 V/m in the atmospheric electrostatic potential while it remained in the water, and also was able to distinguish the direction of this atmospheric potential (Brown, 1962a). The activity of *Drosophila* was reduced by sudden exposure to an electrostatic potential gradient of 1 - 6.25 V/m and was independent of the polarity (Edwards, 1960). A 300 V/m horizontal electrostatic field affected *Drosophila* in terms of their turning, activity, and lunar rhythms (Picton, 1965; 1966).

Milburn cites other examples of electric field behavioral sensitivity observed in insects, fish and mammals.

Evidence for magnetic field behavioral reactions cited are from experiments conducted in flies (Becker and Speck, 1964; Picton, 1965; 1966), in honeybees (Lindauer and Martin, 1968), in snails and planarians (Barnwell and Brown, 1964; Brown and Park, 1967), etc. Studies concerning bird navigation and the earth's magnetic field are also reviewed, as well as some experiments correlating human responses with atmospheric changes in the magnetic field. A review of Tromp's research on the dowsing phenomenon is presented.

7.24 REGENERATION

Sisken and Smith, 1975

Very low level DC currents conducted through the ganglia of embryos developing *in vitro* altered the growth of this nervous tissue. A battery and series resistance provided the current to two platinum electrodes designed to create the non-uniform field necessary for these effects. At a current level of 11.5 mA/m^2 for 12 hours, exposed cultures showed "a far greater preponderance" of nerve fibers compared with the control cultures. The data were quantified by scoring the numbers of healthy and dead cultures. Of 44 cultures exposed to currents between 11.5 mA/m^2 and 1.15 $\mu A/m^2$, 41 received the health score, while in the control group of 28 cultures, only 15 were healthy and 13 were dead. To quote the authors: "This increased survival of neurons within the original explant appears to be the result of electrical stimulation." The neurons from the exposed group showed increased differentiation, which is an unusual finding in ordinary culture experiments.

During the exposure periods, which varied from 2 to 74 hours, with most of the exposures in the range of 24 - 50 hours, time lapse photography showed that cell migration and mitotic behavior were stimulated by the field. There was more cell growth in the exposed cultures, and they grew elongated toward the cathode, at the rate of 2.4 mm/day.

The authors discuss their findings in view of chemical means of producing similar effects, such as use of nerve growth factor (NGF) or cyclic adenosine monophosphate (cyclic AMP). Stimulated growth in the electric field, they speculate, may be due to direct stimulation or may proceed by an alteration in the calcium ion concentration, causing effects on the cell membrane.

Smith, 1974

If the foreleg of a frog (*Rana pipiens*) is amputated, it would not normally regenerate, but the use of a small DC electrical current (1×10^{-7} A generated by a battery and series resistance) results in true regeneration of the limb. The right forelimb of a frog is amputated with a scissors at the mid-forearm level and then the electrical circuit is surgically implanted with the positive terminal located beneath the skin on the back and the negative lead (insulated) brought to the end of the stump. The electrical current flows for 1 month, after which time the circuit is removed, although regeneration continues for some time thereafter.

The degree of regeneration depends upon current level (see Becker, 1972c below or earlier work by Smith) and the exact position in the stump to which the negative lead is brought. In these experiments, Smith grades the animal's response on a scale of 1 - 4, in which Grade 1 shows essentially no regeneration; Grade 2 some outgrowth past the stump with cartilage formed beyond the usual callus; Grade 3 some outgrowth and the development of structure that is well organized; and Grade 4 structure that is very well organized, with the presence of at least 1 well-formed, movable digit. Depending upon the location of the negative lead at the stump, Smith was able to obtain as much as 24% Grade 4 regeneration. Micrographs demonstrate that the tissues growing past the original amputation line may be well-organized and indistinguishable from the normal tissue development that occurs embryonically. In Smith's experience the electrical stimulation must be given within 36 hours of the amputation for successful regeneration, which may proceed for up to 9 months following the 1-month-long electrical treatment.

Smith discusses two possibilities to explain these remarkable results: 1) enhancement of neurite outgrowth; or 2) the transfer of "information-bearing molecules" by electrophoresis. He also considers the possibility of a direct effect on the tissues but, lacking facts, he declines further speculation. The importance of position in stimulating regeneration may be evidence that during embryonic growth there are electrical patterns which govern development and that during regeneration, correct lead placement mimics the embryonic state and "re-awakens long-dormant potential of adult limb tissues." There is evidence from embryology that polarization plays an important role in the early stages of embryonic development, and speculation that it remains an important parameter throughout the development process.

Becker, 1972c; Becker and Spadaro, 1972

In this experiment, 21-day-old male rats were subjected to amputations to determine the effectiveness of various currents in producing regeneration. The experimental animals were surgically implanted with a bimetallic (silver-platinum) element which produced between 3 and 6 x 10^{-9} A when in series with a resistor. The control animals either had no implanted device, an implanted device yielding very low currents (10^{-12} - 10^{-10} A) or one yielding very high currents (5 - 15 x 10^{-9} A).

The results, demonstrated by photomicrographs taken at various times after implantation, show that for the controls without an electrical device, simple wound repair with some bone growth occurred and similarly, for the low current controls, very slight growth was seen. In the high-current controls, bone resorption occurred and cysts formed. The experimental animals showed substantial regeneration, even to the appearance, in 8 of the 22 animals, of "an organized multi-tissue structure with anatomical characteristics appropriate to the morphological area," that is, the growth of muscle, bone and connective tissues. Of the 8 well-organized regenerative growths, 2 were visible at 7 days and the remainder on sacrifice of the animals at later times up to 28 days. It should be noted that some of the regenerates developed abnormal structures, such as a supernumerary humerus which, in itself, was correctly formed.

Although the regenerative growth shown in frogs (amphibians) is remarkable, Becker's use of a mammal in these experiments is still more notable. Becker mentions possible muscle regeneration previously reported in rats, but the regeneration shown here involving a multitissue extremity is unique. Regeneration has also been seen in the newborn oppossum, but because a newborn oppossum is more like a fetus than the newborn of other mammals, it is not quite so remarkable as Becker's results.

Becker feels that there is a control system regulating growth and development, the "key element of which is the induction of blastema formation in response to appropriate electrical factors."*

7.25 ELECTRIC FIELD EFFECTS ON AMOEBA

Friend, 1972; Friend et al., 1975

The giant amoeba, *Chaos chaos*, was observed under the influence of a uniform electric field produced between two parallel platinum wires placed in the medium containing the amoeba. The changes in shape, orientation, and motility which occur depend on field strength and frequency. Investigations over a range from below 1 Hz to about 100 kHz show that at frequencies below 100 Hz the pseudopodia are extended perpendicular to the imposed AC field but not to the extreme achieved at frequencies between 100 Hz and 1 kHz, in which case the pseudopodia parallel to the field are withdrawn, and only perpendicular pseudopodia remain. At higher frequencies (between 1 and 100 kHz) the amoeba loses its pseudopodia and assumes an elliptical shape perpendicular to the field, while at still higher frequencies (above 100 kHz) the elliptical amoeba orients itself parallel to the field. The low frequency changes occur within 10 - 15 minutes at field strengths below 10^3 V/m. The initial effects observed in the ELF

---

* A blastema is a cellular matrix which, in the embryo, generates the organ.

range occur at a few hundred volts per meter, and involve an inward drift of cytoplasm parallel to the field. Field strengths of up to 1.5 kV/m cause no apparent damage to the ameoba, which returned to apparent normalcy within 5 or 10 minutes after the field is removed.

Two mechanisms causing these effects are suggested. The first, an effect on the ion exchange properties of the cell membrane, may be responsible for localized contractions as postulated in a model for DC effects on amoeba. Secondly, dielectric forces (dielectrophoresis, etc.) may play an important role in the observed changes by the imposition of physical forces on the amoeba.

At "higher field levels" (presumably above 10^3 V/m) frequency-dependent disturbances in the membrane are observed. Blisters called "blebs" appear on the amoeba and eventually detach themselves or break down with repair of the membrane following rapidly. Eventually more blebs appear until the membrane disintegrates. Also, pseudopodia may be pinched off in response to high values of the electric field at low frequencies. An interesting characteristic of bleb formation is that there seem to be areas of the membrane which favor bleb formation.

Friend (1972) presents a detailed electrical model of the amoeba, which may be extended to other cells, from which he draws a number of interesting conclusions about the interaction between external currents and the polysaccharide coat outside of the cell membrane proper. From these considerations, he states: "It seems clear that the primary site of interaction for low frequencies must be at or near the surface." He indicates that there is a continuum of effects between DC, ELF, and higher frequencies and that effects merely due to physical forces must be contrasted with those that have a more complex origin in the interaction of the field with cell mechanisms. In this discussion he states that it is possible that "cytoplasmic withdrawal or other more subtle cytoplasmic phenomena may occur even at levels significantly below 500 V/m at ELF, perhaps even at field levels of a small fraction of 100 V/m when applied chronically."

Discussing the usefulness of extrapolation of these results on the amoeba to other life forms, Friend points out that objects subjected to an electric field in the air would have internal fields of much lower amplitide than in the case of an amoeba subjected to fields in a liquid medium. Second, he points out that while Chaos is a useful model, it differs in many ways from a mammalian cell, making scaling to other cell types unwise.

7.26 REFERENCES

Adey, W.R. 1975. Introduction: Effects of electromagnetic radiation on the nervous system. Ann. N.Y. Acad. Sci. 247: 15.

Barnwell, F.H. and Brown, F.A. 1964. "Response of Planarians and Snails." IN: "Biological Effects of Magnetic Fields," M. Barnothy, ed. Plenum Press, New York, p. 263.

Bawin, S.M., Gavalas-Medici, R.J. and Adey, W.R. 1973. Effects of modulated very high frequency fields on specific brain rhythms in cats. Brain Research 58: 365.

Becker, G. and Speck, U. 1964. Untersuchungen uber die magnetifeldorientierung von dipteren (Magnetic field orientation in Diptera). Zeits. ver. Physiologie 49: 301.

Becker, R.O. 1972c. Stimulation of partial limb regeneration in rats. Nature 235: 109.

Becker, R.O. and Spadaro, J.A. 1972. Electrical stimulation of partial limb regeneration in mammals. Bull. N.Y. Acad. Med., 2nd Series 48: 627.

Beischer, D.E., Grissett, J.D. and Mitchell, R.D. 1973. "Exposure of Man to Magnetic Fields Alternating at Extremely Low Frequency." Naval Aerospace Medical Research Laboratory, Pensacola, Fla. Rept. NAMRL-1180. Available from NTIS as AD 770140.†

Beischer, D.E. and Knepton, J.C. 1966. "The Electroencephalogram of the Squirrel Monkey (Saimiri sciureus) in a Very High Magnetic Field." NAMI-972, Naval Aerospace Medical Institute, Naval Aerospace Medical Research Laboratory, Naval Air Station, Pensacola, Fla.

Belyaeva, P.I. and Stroikova, K.V. 1975. "The Physiological Investigation of the Effects of a High Tension, Industrial Frequency Electrical Field on Living Organisms." Abstract by G. Knickerbocker available in Special Publ. #10, IEEE Power Engineering Society, Piscataway, N.J.(1975).

Blakemore, R. 1975. Magnetotactic bacteria. Science 190: 377.

Blanchi, D. et al. 1973b. Exposure of mammalians to strong 50 Hz electric fields: 2) effect on heart's and brain's electrical activity. Arch. Fisiol. 70: 33.

Brown, F.A. 1962a. Response of the planarian, Dugesia, to very weak horizontal electrostatic fields. Biol. Bull. 123: 282.

Brown, F.A. and Park, Y.H. 1967. Association formation between photic and subtle geophysical stimulus patterns--a new biological concept. Biol. Bull. 132: 311.

Bullock, T.H. 1973. Seeing the world through a new sense: electroreception in fish. Am. Sci. 61: 316.

Coate, W.B. et al. 1970. "Project Biological Effects Test Program. Pilot Studies. Final Report." Hazleton Laboratories, Inc. Available from NTIS as AD 717408.†

deLorge, J. 1972. "Operant Behavior of Rhesus Monkeys in the Presence of Extremely Low Frequency-Low Intensity Magnetic and Electric Fields: Experiment 1." Naval Aerospace Medical Research Lab., Pensacola, Fla. Rept. NAMRL-1155. Available from NTIS as AD 754058.†

deLorge, J. 1973a. "Operant Behavior of Rhesus Monkeys in the Presence of Extremely Low Frequency-Low Intensity Magnetic and Electric Fields: Experiment 2." Naval Aerospace Medical Research Lab., Pensacola, Fla. Rept. NAMRL-1179. Available from NTIS as AD 764532.†

deLorge, J. 1973b. "Operant Behavior of Rhesus Monkeys in the Presence of Extremely Low Frequency-Low Intensity Magnetic and Electric Fields: Experiment 3." Naval Aerospace Medical Research Lab., Pensacola, Fla. Rept. NAMRL-1196. Available from NTIS as AD 774106.†

deLorge, J.1974. "A Psychobiological Study of Rhesus Monkeys Exposed to Extremely Low Frequency-Low Intensity Magnetic Fields." Naval Aerospace Medical Research Lab., Pensacola, Fla. Rept. NAMRL-1203. Available from NTIS as AD 000078.

Durfee, W.K. et al. 1975a. "Extremely Low Frequency Electric and Magnetic Fields in Domestic Birds." University of Rhode Island, Kingston, R.I.†

Durfee, W.K., Martin, P.J. and Muthukrishnan, S. 1975b. "The Influence of a 5 Gauss, 60 Hz Magnetic Field upon Growth and Development of the Chick Embryo." Dept. of Animal Sciences, University of Rhode Island, Kingston, R.I. †

Edwards, D.K. 1960. Effects of articially produced atmospheric electrical fields on the activity of some adult Diptera. Can. J. Zool. 38: 899.

Friend, A.W. 1972. "Low Frequencies, Motile Cells, Measurements and Models. Part I." Rept. No. 4, Naval Medical Research Inst., Bethesda, Md.†

Friend, A.W., Finch, E.D. and Schwan, H.P. 1975. Low frequency electric field induced changes in the shape and motility of amoebas. Science 187: 357.

Gavalas, R.J. *et al*. 1970. Effect of low-level, low-frequency electric fields on EEG and behavior in *Macaca nemestrina*. *Brain Res*. *18*: 491.

Gavalas-Medici, R. and Magdaleno, S. 1975. "An Evaluation of Possible Effects of 45 Hz, 60 Hz and 75 Hz Electric Fields on Neurophysiology and Behavior of Monkeys." Univ. of California, Los Angeles. ONR Technical Paper.†

Gengerelli, J.A. and Holter, N.J. 1941. Experiments on stimulation of nerves by alternating electrical fields. *Proc. Soc. Exp. Biol. Med*. *46*: 532.

Goodman, E.M., Greenebaum, B. and Marron, M.T. 1975. "Effects of Extremely Low Frequency Electromagnetic Fields on Growth and Differentiation of *Physarum polycephalum*." Univ. of Wisconsin, Parkside, Wis. Technical Report Phase I (Continuous Wave)

Graue, L.C. 1974. "Orientation of Homing Pigeons (*Columbia livia*) Exposed to Electromagnetic Fields at Project Sanguine's Wisconsin Test Facility." Bowling Green State Univ. Final Reprt. Office of Naval Research, Dept. of the Navy. Available from NTIS as AD 780948.†

Grissett, J.D. and deLorge, J. 1971. "Central-Nervous-System Effects as Measured by Reaction Time in Squirrel Monkeys Exposed for Short Periods to Extremely Low-Frequency Magnetic Fields." Naval Aerospace Medical Research Lab., Pensacola, Fla. Rept. NAMRL-1137. Available from NTIS as AD 731994.†

Hopkins, C.D. 1974. Electric communication in fish. *Am. Sci*. *62*:4

Jaffe, L.F. 1966. Electrical currents through the developing fucus egg. *Proc. Nat. Acad. Sci*. *56*: 1102.

Knickerbocker, G.E., Kouwenhoven, W.B. and Barnes, H.C. 1967. "Exposure of Mice to a Strong AC Electric Field-An Experimental Study." IEEE Trans. Pow. App. Sys. PAS-86, No. 4.

Kouwenhoven, W.B. *et al*. 1966. Body currents in live line working. *IEEE Trans. Pow. App. Sys*. PAS-85, No. 4.

Krueger, W.F. *et al*. 1972. Influence of low-level electric and magnetic fields on the growth of young chickens. *Biomed. Sci. Instrum*. *9*: 183.

Lindauer, M. and Martin, H. 1968. The earth's magnetic field affects the orientation of honeybees in the gravity field. *Zeits. ver. Physiologie* *60*: 219.

Lott, J.R. and McCain, H.B. 1973. Some effects of continuous and pulsating electric fields on brain wave activity in rats. *Int. J. Biometeor*. *17*: 221.

Marr, M.J., Rivers, W.K. and Burns, C.P. 1973. "The Effect of Low Energy, Extremely Low Frequency (ELF) Electromagnetic Radiation on Operant Behavior in the Pigeon and the Rat." Georgia Inst. of Techology, Final Report. Available from NTIS as AD 759415.†

Marron, M.T., Goodman, E.M. and Greenbaum, B. 1975. Mitotic delay in the slime mold *Physarum polycephalum* induced by low intensity 60 and 75 Hz electromagnetic fields. *Nature*: *254*: 66.

Milburn, B.B. 1971. "Very Low Frequency Electromagnetic Fields and Behavior." Dissertation, Univ. of Calif. at Los Angeles. Available from University Microfilms, Ann Arbor, Mich. as #72-2866.

Mittler, S. 1972. "Low Frequency Electromagnetic Radiation and Genetic Aberrations." Final Report, Northern Illinois Univ. Available from NTIS as AD 749959.†

Moos, W.S. 1964. A preliminary report on the effects of electric fields on mice. *Aerospace Med*. *35*: 374.

Novak, B. and Bentrup, F.W. 1973. Orientation of fucus egg polarity by electric AC and DC fields. *Biophysik* *9*: 253.

Persinger, M.A. 1974. "ELF and VLF Electromagnetic Field Effects" (ed.), Plenum Press, New York.

Persinger, M.A., Ludwig, H.W. and Ossenkopp, K-P. 1973. Psycho-physiological effects of extremely low frequency electromagnetic fields: a review. *Percept*. *Mot*. *Skills* *36*: 1131.

Persinger, M.A. and Pear, J.J. 1972. Prenatal exposure to an ELF-rotating magnetic field and subsequent increase in conditoned suppression. *Dev*. *Psychobiol*. *5*: 269.

Picton, H.D. 1965. The responses of *Drosophila melanogaster* to weak magnetic fields. *Dissertation Abst*. *25* (11): 6727.

Picton, H.D. 1966. Some responses of *Drosophila* to weak magnetic and electrostatic fields. *Nature* *211*: 303-304.

Robinson, K.R. and Jaffe, L.F. 1975. Polarizing fucoid eggs drive a calcium current through themselves. *Science* *187*: 70.

Rommel, S.A. and McCleave, J.D. 1972. Oceanic electric fields: perception by American eels? *Science* *176*: 1233.

Rommel, S.A. and McCleave, J.D. 1973a. Sensitivity of American eels (*Anguilla rostrata*) and Atlantic salmon (*Salmo salar*) to weak electric and magnetic fields. *J. Fisheries Research Board of Canada* *30*: 657.

Rommel, S.A. and McCleave, J.D. 1973b. Prediction of oceanic electric fields in relation to fish migration. *J. Cons. Int. Explor. Mer.* *35*: 27.

Schua, L. 1953. Die Fluchtreaktion von Goldhamstern aus elektrischen Feldern. (Avoidance reaction of hamsters in an electric field.) *Naturwissenschaften* *40*: 514.

Schua, L. 1954. Die Wirkung von luftelektrischen Feldern auf Tier (The effect of atmospheric electric fields on animals.) Deutsche Zoologische Gesellschaft. *Verhandlungen* *7*: 435.

Sisken, B.F. and Smith, S.D. 1975. The effects of minute direct electrical currents on cultured chick embryo trigeminal ganglia. *J. Embryol. Exp. Morphol.* *33*: 29.

Smith, L.T. 1975. "The Influence of Continuous Early Exposure to an ELF Magnetic Field upon Subsequent Adult Social Behavior of the Domestic Fowl." Dept. of Animal Sciences, Univ. of Rhode Island, Kingston, R.I. †

Smith, S.D. 1974. Effects of electrode placement on stimulation of adult frog limb regeneration. *Ann. N.Y. Acad. Sci*: *238*: 500.

Southern, W.E. 1973. "Orientation Behavior of Ring-Billed Gull Chicks (*Larus delawarensis*) Exposed to Project Sanguine's Electri and Magentic Fields." Northern Illinois Univ. Final Report.†

Southern, W.E. 1975. Orientation of gull chicks exposed to Project Sanguine's electromagnetic field. *Science* *189*: 143.

Straub, K.D., Kendrick, J.Z. and Jackson, H. 1972. "Effects of Low Frequency Electrical Current on Various Marine Animals." Naval Air Development Center (NADC), *72126-Cs*. Available from NTIS as *AD 749335*.†

Terzuolo, C.A. and Bullock, T.H. 1956. Measurement of imposed voltage gradient adequate to modulate neuronal firing. *Proc. N.A.S.* *42*: 687.

Walcott, C. 1974. The homing of pigeons. *Am. Sci.* *62*: 542.

Webb, H.M., Brown, F.A. and Brett, W.J. 1959. Fluctuations in rate of locomotion in *Ilyanassa*. *Biol. Bull.* *117*: 431.

Webb, H.M., Brown, F.A. and Schroeder, T.E. 1961. Organismic responses to differences in weak horizontal electrostatic fields. *Biol. Bull.* *121*: 413.

Weisenseel, M.H., Nuccitelli, R. and Jaffe, L. 1975. Large electrical currents traverse growing pollen tubes. *J. Cell. Biol.* *66*: 556.

Yates, V. *et al*. 1975. "The Influence of ELF Magnetic and Electric Fields upon the *in vitro* Growth Rate of Chicken Embryo cells." Dept. of Animal Pathology, Univ. of Rhode Island, Kingston, R.I.†

Chapter 8

SPECIAL TOPICS

8.1 CELLS IN CONTACT: ELECTRICAL PROPERTIES

Loewenstein and Kanno, 1964; Wiener et al., 1964; Loewenstein, 1966; Dewey and Barr, 1962; Azarnia et al., 1974; DiCaprio et al., 1975; Garfield and Daniel, 1974; Spanswick, 1972

The advent of high resolution electron microscopy reveals that there are regions in smooth muscle tissues where cells form intimate contacts in which the cell membranes make, in some cases, an intimate three-layer sandwich by the union of the two outer outer membrane surfaces (Dewey and Barr, 1962). This junction, called a nexus, was examined from the point of view of membrane transport phenomena (Loewenstein and Kanno, 1964) in the case of a gland cell junction similar to the muscle nexus. Examination of the membrane permeability disclosed that there was free movement of ions from cell to cell, although there still was a very high resistance to the motion of ions from a cell to the extracellular medium.

It was also demonstrated that the electrical resistance across cell contact membranes is much lower than the resistance of the surface membrane, indicating preferred pathways for electronic and ionic currents between cells. It is possible to allow currents to flow from cell to cell, effectively bypassing the cell exterior, using the cytoplasm and contact surfaces for a path of current continuity. Using the methods of transmission line analysis, the voltage variation along the length of the current path can be calculated and is in good agreement with the experimental data (Loewenstein and Kanno, 1964).

Electron micrographs (Wiener et al., 1964),disclosing the complex structures at the regions of cell contact which are responsible for the electrical and ionic properties, have been made from salivary gland tissue, renal and liver cells, bladder cells, nerve cells, smooth muscle and cardiac muscle. Indeed, cell communication through contact regions seems to be a general phenomenon of cell association and particular contacts known as gap junctions, tight junctions, etc. have been distinguished. Certain nerve and striated muscle cells lack the ability to communicate by contact and are surrounded by wide fluid spaces preventing electrical "cellular crosstalk."

Cell communication can be stopped (functional uncoupling) by removal of the Ca^{++} ions, which has a reversible effect on membrane permeability but an irreversible effect on cell coupling (Loewenstein, 1966). Uncoupling may also be effected by enzymes (trypsin) or by the use of a hypertonic solution to produce mechanical forces. Loewenstein suggests that cellular interconnections are a means for tissue regulation in growth and differentiation, pointing out that cancer cells of the liver lack intercellular communication.

Garfield and Daniel (1974) found that the cell contacts in smooth muscle from the uterus of a rat had no nexuses but did have a variety of intermediate and close contacts which apparently allow the uterine muscle to act as an integrated organ in contraction.

The role of intercellular communication in the very early stages of embryo development is being investigated by DiCaprio *et al.* (1975), who discovered that, at least to the stage of 8 cells in the embryo of *Xenopus*, there is complete electrical communication between all the cells. A microscopic examination shows that there is extensive intercellular contact provided by the modeling of the surface features on the adjoining cell faces. The authors conclude that "the high order of connectivity therefore supports the view that intercellular junctions are significant in the normal development of the embryo." They also suggest that the connections may be more important for molecular flow than electrical flow.

Spanswick (1972) discusses electrical coupling in plant cells (*Elodea*) in terms of the role of intercellular ionic transport and intercellular communication.

8.2 ELECTRICAL CATARACTS

Cataracts commonly occur as a consequence of serious electrical injuries to the head such as from contact with high voltage power lines or from lightning strikes (Hanna and Fraunfelder, 1975; Fraunfelder and Hanna, 1972). The lens opacity may develop over a varying period of time. The speed with which the first symptom develops (formation of vacuoles in the lens) is related to the amount of electrical energy supplied to the head. Experiments in rabbits have shown that 25 joules to the eye globe causes the formation of vacuoles within 30 days while shocks of 50 to 300 joules result in the formation of vacuoles within 2 or 3 days. Following vacuole formation, there is a variable period of months to years during which the epithelial cells undergo changes which suggest attempts to differentiate into lens fibers and which eventually result in serious opacities of the lens.

No cataracts are known from exposures to ELF electrical fields which do not, of course, result in such large deposits of energy to the eye.

8.3 AIR IONIZATION

The biological effects of the ion content in air have been studied in the context of meteorological effects on human behavior and in laboratory studies of behavior or biochemical response. Krueger et al. (1968) have advanced the hypothesis that positively ionized air results in a small concentration of oxonium $[H^+(H_2O)]$ or hydronium $[(H_3O)^+(H_2O)_n]$ ions and negatively ionized air results in the species $O_2^-(H_2O)_n$, $OH^-(H_2O)_n$, CO_3^-, and CO_4^-. In controlled laboratory studies, they have been able to detect changes in the blood levels of 5-hydroxytryptamine (a neurally active hormone also known as serotonin) at ion concentrations of $4 - 5 \times 10^5$ ions/cm^3 (about 1 part per 7×10^{13} molecules of air). The positive ions were observed to increase the serotonin levels in mice 24 hours after the start of the exposure and the negative ions lowered the serotonin levels in a similar period of time. Some refinements have been made in experiments in this field, but the results are still viewed skeptically (Krueger et al., 1968). At present, there is no theoretical understanding of the effects, which would require a biochemical mechanism affected by ion concentrations of less than 1 part in a trillion.

Nonetheless, we can enumerate several examples of the effects attributed to alterations of the air's ion concentration:

1) Decreased growth rates were observed in heart cell cultures exposed to positive ions and in fibroblast cultures also exposed to positive ions. Fibroblasts exposed to the negative ions had augmented growth rates (Worden, 1961).

2) More gastric ulcers were observed in rats exposed to combined stress (starvation) and positive ions than in the control rats, and the rats exposed to the stress and the negative ions had fewer ulcers (Cupcea et al., 1959).

3) Both ion species affect growth rates in plants (Krueger et al., 1968) and accelerate development of silkworm eggs and larvae.

4) Both ion species reportedly affect the visual reaction time in humans, resulting in randomly distributed increases or decreases in the reaction times (Knoll et al., 1964).

See Table 8-1 for a listing of other biological effects attributed to ions in the air.

Field studies find that the ozone produced by corona on a transmission line is at concentrations below the limit of detectability, which is an increment of about one part per billion above the background.

A very few experiments on electric field effects have recognized the possibility that the air ions created under laboratory conditions might be a confusing variable in the experiment (Hauf, 1974; Johansson _et al._, 1973) but, at present, there is little evidence that this has been the case.

TABLE 8-1. SOME BIOLOGICAL EFFECTS OF AIR IONS*

	Negative Ions	Positive Ions
Animal Subjects		
Blood pH	Marked Increase	Small Decrease
Growth:		
Embryonic cells	Enhanced	Depressed
Tissue cells	Enhanced	Depressed
Newborn litters	Enhanced	None
Bacteria	Inhibited	Inhibited
Tracheal Ciliary Activity	Increased	Depressed
Vasomotor Activity	Dilation	Constriction
Human Subjects		
Performance	Improved	Decreased
Work Capacity	Increased	-
Disposition	Cheerful	Depressed
Reaction Time	Decreased	Decreased
Equilibrium	Improved	-
Burn Recovery and Healing	Enhanced	-
Pain	Relieved	-
Allergic Disorders	Relieved	-
Vitamin Metabolism	Enhanced	-

* Beal, 1974

8.4 SURVEYS OF FARMERS AND EXPOSED PERSONS

Busby et al., 1974

Eighteen farms whose land is crossed by a 765 kV transmission line were surveyed to find out if their farm operations were affected by the introduction of the line.

Twelve farmers (producing corn, wheat, soybeans, beef, or hogs) were in a region in Ohio where the line was energized to 720 kV for one crop year. No electric field effects were reported by the farmers. Two of the dairy farmers reported increased milk production, two less milk production. None of the farmers reported changes in the animals' grazing habits; seven farmers felt their yields were affected by the construction of the lines; three made note of slight shocks they had experienced; and one of a bad shock. Three farmers reported that animals seemed to prefer resting under the towers.

Ware, 1975

One hundred and twenty-five farmers grazing animals on lands near a 765 kV transmission line were surveyed. Of these, 116 reported no problems relative to the line; seven reported incidents of shock; two noted no problems but did say that the animals don't graze under the line; one person felt that an animal had been killed by the line's presence, but a consulting veterinarian found that this was not a reasonable assumption for the cause of the animal's death.

Strumza, 1970

Possible evidence for health problems was sought by a comparison of the consumption of medical and pharmaceutical services by persons living near to or far from high voltage lines (in the range of 200 to 400 kV). The data consisted of the number of medical office visits, home visits, consultations with specialists, and pharmaceutical bills for employees of the French electric company (E.D.F.) living and working in various rural locations throughout France. Two hundred and sixty-seven persons were classified as exposed because they lived within 25 meters of a high voltage transmission line, and 258 formed the class of non-exposed persons, for they lived and worked 125 meters or more from the lines. The average length of exposure was three years, nine months.

A statistical analysis of the data showed no significant effects in any categories excepting specialists' consultations among children, which was explained by the presence of two children in the unexposed group who made an exceptionally large number of visits to a specialist. The authors point out that the sample populations were matched according to social class, job experience and region where they lived. The data and statistical parameters are reported in detail.

8.5 ELECTRIC FIELD EFFECTS ON PLANT LIFE

Several studies illustrate the use of relatively strong DC electric fields to affect the viability of seeds during storage and seed germination, or to influence the growth of plants. Pammenter _et al_. (1974) reported that stored seeds subjected to an artificial aging process progressively lose the ability to germinate, so that after 30 days of treatment only 10% of the seeds are viable. However, by placing the seeds on an electrode at 300 volts the germination rate after the 30-day aging treatment was improved to about 70%. The data demonstrate that a substantial improvement in seed viability occurs for all degrees of aging and that the protected seeds also show a lower percentage of chromosomal abnormalities. These effects are attributed to a form of "cathodic protection" in that the electrode provides electrons to react with the free radicals and thus inhibit the normal course of lipid peroxidation by which, it is believed, the cell membrane is destroyed in both artificial and natural aging processes.

Sidaway (1968) examined the germination of seeds in an electrostatic field of 36 kV/m and came to the conclusion that a field with the positive pole above the seeds inhibits germination and, to a lesser extent, the negative pole above improves the rate of germination.

Murr (1964; 1965) studied the destructive effects which occur in plants grown in fields of 20 - 80 kV/m (DC) created by a plate located above the ground in which the plants were rooted. The leafy tips of grass plants were injured or killed by this strong electrostatic field, which was intensified to much higher surface field values at the leaf tip. From his studies, Murr finds a surface field of 110 - 130 kV/m is about the threshold for destructive effects in these plants. Murr (1964) speculates that the mechanism by which the plants are injured involves an acceleration of enzyme activity leading to impaired cell respiration and further destructive events.

M. W. Miller (1974) exposed the roots of the mung bean plant (_Vicia faba_) to electric or magnetic fields at 75 Hz wich no indication of an effect on either growth or mitotic index. No chromosomal anomalies were seen. The electric field exposure was made in a plastic tank filled with a plant nutrient solution in which there was a current of 0.1 A/m^2 and an electric field of 10 V/m. The control plants were in a similar tank for which the electrodes were not energized. The magnetic field experiments were conducted at 0.5, 5, and 17 G in plastic cylinders containing the plants in a liquid nutrient medium and wrapped with electrical wire. The authors cite other "well-documented" research (Beischer, 1965; Audus and Whish, 1964; Dunlop and Schmidt, 1969; Mericle _et al_., 1964) at higher field strengths (from several hundred to several thousand gauss) which did find effects on plant growth and the mitotic index, but in view of the very much smaller magnetic fields employed here, there is no inconsistency.

A survey of plant growth was made in 1970 at the site of the first Project Sanguine test transmitter used in 1962, and at locations near telephone lines, power lines, and at the Wisconsin Test Facility antenna (McCormick _et al_., 1971). No effects were seen in the plants, their populations, or species distributions which could be attributed to the exposure to the ELF antenna fields (whose strength is unspecified here). Altered death rates observed among the buds of pine trees located near the antenna were attributed to the effects of the tree cutting in the immediate environment. This was confirmed by observation of pine buds in cut areas remote from the antenna.

Our review of the research on the effects in plants is incomplete and there are many other studies which may be of interest.

8.6 MAGNETIC FIELDS IN THE HUMAN BODY

Cohen, 1973; 1975a; 1975b

Cohen has pioneered in measuring the magnetic fields produced in the human body by brain rhythms, cardiac electricity activity, or magnetic contamination. The techniques involve the ability to detect very small fields (as low as 1 nG) and the development of a room with a very high degree of magnetic shielding. The fields that Cohen detects using sophisticated solid state devices (the Fluxgate magnetometer or the superconducting quantum interference device, SQUID) are generated in the body by the ion currents associated with muscle or nerve discharges. The static magnetic fields resulting from injuries to heart muscle or as the result of inhalation of magnetic particles (e.g., magnetite from asbestos) into the lungs (see Table 8-2) can also be measured.

Magneto-cardiograms have been measured in the frequency range of 0.1 Hz to 40 Hz and at DC. The magneto-encephalogram (at a level of a few tens of nG) demonstrates signals similar to the electric encephalogram except for the prominent absence of the theta waves at 5 Hz, which Cohen speculates may be due to the nature of the current generator in the cortex responsible for the theta waves.

Magnetic particles in the lung can be measured by magnetizing the subject in a field of from 300 to 1000 G using an external electromagnetic or permanent magnet to achieve maximal signals. The usefulness of this technique for detecting embedded ferromagnetic particles which settle in the lung as a result of inhalation of mining dusts is obvious. It is expected that this technique may have clinical usefulness in determining lung function, especially in lungs damaged by disease or environmental pollutants. This may be achieved by having the patient inhale

TABLE 8-2. MAGNETIC FIELDS IN THE HUMAN BODY *

1. DC Fields		
from injured heart		3×10^{-7} gauss
abdominal		10^{-6}
magnetite in lungs of asbestos miners		3×10^{-5}
(geomagnetic field)		(0.5)
2. AC Fields		
brain waves (alpha rhythm)		10^{-8}
normal brain in sleep		5×10^{-8}
skeletal muscle	(1 - 100 Hz)	10^{-7}
normal heart	(0.1 - 40 Hz)	2×10^{-6}
(geomagnetic field	0.1 - 100 Hz)	(3×10^{-5})
(urban magnetic noise	0.1 - 100 Hz)	(3×10^{-3})

* Cohen, 1975b

a very small amount of magnetic dust and observing the pattern of its deposition, and the decay of that deposition with time, as a measure of lung function.

8.7 ORIENTATION OF DNA IN A MAGNETIC FIELD

Maret et al., 1975

This experiment used very large fields (above 10,000 G) to demonstrate the possibility of orienting nucleic acids and similar polymer molecules using dilute solutions of macromolecules. This technique allows determination of the "persistence length" of the polymer. This persistence length is a consequence of the flexibility of the polymer chain and its curvature which give a resulting anisotropic diamagnetic susceptibility by a summation over the diamagnetic units in the chain. Using an optical technique to measure the magnetically induced change in the index of refraction (the Cotton-Mouton Effect), the authors were able to achieve high resolution and great sensitivity. Measurements were made in a DNA solution obtained from calf thymus and in two synthetic polymers which, at a magnetic field strength of 14 tesla (140,000 G), were oriented by 1%.* The temperature dependence of the orientation effect showed a loss of orientation at the temperature at which the DNA was denatured (above ∿ 60°C).

Analysis of the data leads to the conclusion that the base pairs are responsible for magnetic alignment, are themselves diamagnetically anisotropic due to their ring structures, and consequently tend to orient themselves parallel to the applied field, resulting in a perpendicular orientation for the DNA filamentary axis.

The authors distinguish between effects in electric and magnetic fields in their summary paragraph, which we quote: "We have shown that in magnetic fields the small but steady orientation of macromolecules like DNA depends only on the

---

* The orientation parameter (β) expresses the degree to which the orientation of a hypothetical rigid rod is non-random in terms of the probabilities of it being completely aligned with one of the three cartesian axes. Letting W_x, W_y and W_z be the probabilities of complete alignment along one of the axes, the orientation parameter is,

$$\beta = \frac{(W_x - W_z)}{W_z} + \frac{(W_y - W_z)}{W_z},$$

which is related to the magnetic susceptibilities parallel and perpendicular to the rod, the field strength and the temperature. (See the reference for details.)

relative position of well-defined diamagnetic parts inside the macromolecule. Furthermore, the macromolecule is always in thermal equilibrium with its ionic solvent. In contrast, the orientation behavior of polydisperse, polyelectrolytic solutions in electric fields normally depends on contributions of constant and induced electric dipole moments and of the hyperpolarizability of the macromolecule. Therefore, in order to avoid heating effects caused by electrolytic currents, transient, non-equilibrium orientation techniques are in use and a spectrum of relaxation times has to be interpreted."

8.8 BIOLOGICAL EFFECTS OF LARGE MAGNETIC FIELDS

Biological effects in the range of 100 G up to very large fields of 100,000 G (10 tesla) or even more have been investigated with increasing interest but, because our primary interest is in magnetic fields commonly found in the environment, rather than fields unique to a few industrial or laboratory situations, we have segregated most of the reports to this section and do not present them in as great detail. However, the "crosstalk" between the investigations at high and low field strengths may be valuable; in addition, the references below could assist in making an evaluation of the effects of large magnetic fields. Some work at very low fields (or "zero field") is also mentioned.

Becker, 1963

Becker reviews the history of magnetic field research up to 1963 discussing phenomena like magnetic phosphenes, in which human subjects observe flashes of light when in a pulsed or AC magnetic field above 100 G at frequencies between 10 and 100 Hz. The maximum effect occurs at about 20 Hz. (For more about magnetic phosphenes, see Barlow *et al.*, 1947 and Valentinuzzi, 1962.) Another AC effect suspected at the time is the inhibition of growth observed in a number of experiments. Because only inhomogeneous DC fields exert a force on a dipole, it seems likely that homogeneous DC fields would be biologically ineffective. Experiments at several thousand G on simple organisms (bacteria, yeast, or fungi) showed no effects with homogeneous fields, although other experiments in mice (Barnothy's research) show that constant, inhomogeneous, magnetic fields interfered with the white blood cell formation. Mice exposed for 10 days at 5500 G were killed and effects on wound healing were observed at 10^5 G.

Inhomogeneous DC fields at 4 G were reported to cause reduced growth and 1 G fields altered biological rhythms. At 10^3 G, tumor takes in experimental animals were lowered and there was some demonstration of "prophylaxis" against subsequent tumor development. Becker reports a plant experiment in which a dose-response relationship is shown for inhomogeneous fields.

At 10^4 G, some tissue cultures were unaffected, while tumor cell cultures were killed. At 45,000 G there were no fatal effects during a 1-hour exposure of mice, but *Drosophila* were 100% destroyed and sea urchin eggs affected.

Experiments on neural transmission or changes in threshold neuronal firing showed no effects with homogeneous magnetic fields. Evidence suggesting electronic (rather than ionic) conduction in nervous tissue arises from experiments done at 10^4 G.

Finally, the influence of magnetic fields on birds, migrations, biorhythms, or human behavior (demonstrated by rates of mental illness or suicide) are discussed in terms of research known at the time.

Barnothy, 1964; 1969

These two volumes collect many articles on the biological effects of magnetic fields, usually inhomogeneous fields in the range of several thousand G, acting on diverse biological systems. Among the results noted are the inhibition of bacterial growth during the log phase caused by inhomogeneous fields and no lethal effect on *Drosophila* (see Gerencser *et al*., 1964 and Beischer, 1964). Neurath (1964) discussed models for the magnetic interaction, demonstrating that it takes very large (more than 10^5 G) magnetic fields to introduce orientation in a suspension of ellipsoidal cells, e.g., red blood cells. He also demonstrated that because of the low conductivity of biological materials, the Hall effect potentials are too small to be biologically effective, although the motion of animals in fields of 10^3 G induces electric fields that are of the order of 1 V/m.

M. F. Barnothy reported on the higher activity level observed for mice exposed continuously to a 4200 G inhomogeneous field, accompanied by a lowered food use, implying better metabolic use of food. Changes in liver, spleen, adrenal glands and bone marrow were observed after 120 days.

J. M. Barnothy observed differences in the development of mice growing up in large magnetic fields, the exposed mice being smaller. Using homo-transplanted tumors, he observed a smaller rate of tumor growth and eventual rejection of the tumor mass in mice exposed at 4000 G. Barnothy suggests that the rejection in the host occurs as a part of an immune response in the host rather than a magnetic field effect on the cancer cells, which he shows in a further experiment in which animals bearing iso-transplanted tumors (not transplanted from another animal) were similarly exposed to magnetic fields with the result of somewhat increased longevity without rejection. Additionally, magnetic fields of several thousand gauss result in a variation in white blood cell counts, perhaps due to a shortening of the lifespan of the circulating WBCs. In one experiment leukocytosis, induced by the body's response to the effect on the WBCs, was apparently the reason for greater survivability among animals exposed to ionizing radiation (750 R). Normally, at such doses the animals die by effects related to the loss of WBCs.

Constant magnetic fields are reported to affect wound healing by a delay of fibrosis and fibroblast proliferation, but there is no effect on tissue regeneration itself (Gross and Smith, 1964).

There is a discussion (Mulay and Mulay, 1964) of experiments attempting to discover direct effects of magnetic fields on chemical reactions such as redox reactions involving a paramagnetic ion (e.g., the reduction of ferric chloride). The authors point out that the energy of interaction between biological materials and very large magnetic fields is very small, 5×10^3 ergs/mole. Hence, it is suggested that effects may occur by preferential orientation related to currents arising from "a circulation of the pi electrons in certain molecules." The magnetic susceptibility of molecules -- or portions of large molecules -- may be sensitive to the pi electron distribution and thus small changes in magnetic susceptibility may have an influence on the progress of chemical (enzyme) reactions.

Plant response to magnetic fields is said to include acceleration of seed germination, inhibition of seed germination, accelerated growth, and tropisms (curvature away from the field-center). A number of experimental results showing no effects are also reported.

Effects on the central nervous system (Kholodov, 1964) include: intensified visual images for persons under the influence of hypnosis or mescaline, but no similar intensification for normal persons; behavioral conditioning in fish; EEG changes in the rabbit; overall inhibition of the central nervous system; and excitation of the motor system.

Exposures at exceptionally large magnetic fields (140,000 G) disclosed no acute effects on mice, fruit flies, or bacteria, but the early cleavage of sea urchin eggs, shortly after fertilization, was slowed down by a 2-hour exposure to the magnetic field (Beischer, 1964).

Among other reports included is research on enzyme reaction rates (Cook and Smith, 1964; Wiley _et al_., 1964) and the response of planarians and snails to very weak magnetic fields (1 G) (Barnwell and Brown, 1964).

The first volume concludes with a discussion on dowsing -- the human perception of underground discontinuities, e.g., water -- which reports that the dowsing phenomenon is really a detection of magnetic anomalies with the detection threshold at a gradient of about 10^{-4} G/m, which is created when the dowser moves through the magnetic field anomaly. The possibility that birds use magnetic fields as a navigational tool is discussed and a theory which presumes detection of an induced electrical field caused by the bird's motion is presented.

The second volume includes an extensive table listing about 25 experiments on organisms from algae to man at low magnetic field strengths (usually less than 1 G) demonstrating behavioral effects and effects on growth, enzyme activity, and the nervous system.

The effects of extremely low magnetic field environments (less than 8×10^{-4} G) were investigated showing depressed enzyme activity in mouse tissues isolated from the normal geomagnetic field of 0.5 G. These experiments indicate effects only in whole animal exposures for a prolonged period of time which were not reproducible by *in vitro* chemical experiments performed in similar magnetic field conditions. A modern experiment, employing radioactive labeling, shows the rate of DNA synthesis and respiration in tumor cells is depressed about 20% by an exposure of 1 - 3 hours in a DC field at 7300 G (D'Souza *et al.*, 1969).

A direct magnetic field effect on the nervous system is demonstrated in the exposure of an excised frog's heart to large magnetic fields (about 4000 G with a gradient of 8.3×10^4 G/m). Changes in heart rhythm and decreased contraction strength indicate an effect on acetylcholinesterase, the enzyme necessary for nerve transmission. In this and other experiments, Young (1969) and Young and Gofman (1965) give additional evidence for a possible interaction of the magnetic field with acetylcholinesterase which results in a higher rate of hydrolysis, perhaps due to changes in the substrate molecules.

The effects of a magnetic field on the histology of organs taken from mice exposed for long periods of time are reported, with changes of a pathological nature observed at large magnetic fields (DC). The authors (Barnothy and Sumegi, 1969) suggest that the pattern of changes indicates that the magnetic field created a hormone imbalance because the body perceived the magnetic field as an unfavorable stimulus, in a manner not quite consistent with Selye's theory of non-specific stress (the general adaptation syndrome).

Other experiments include measurements of central nervous system reaction, erythrocyte sedimentation rate, white blood cell counts, embryonic development, mutagenesis in *Drosophila* at at low field strengths, inhibition of the neuronal rate of firing, food and activity behavior of mice, and changes in the EKG caused by magnetic fields of 40,000 - 100,000 G. The second volume concludes with a discussion of the magnetic susceptibility of biological tissues, and also presents magnetic susceptibility data for a large number of iron-containing biological molecules and various other organic compounds (the nucleic acids and some carcinogenic and chemo-therapeutic compounds). These data do no immediately indicate biomagnetic mechanisms for biological susceptibility to magnetic fields, but the observation that biochemical compounds may be both paramagnetic and diamagnetic suggests subtle effects which could show up in enzyme reactions.

Biomagnetics seems to hold a more central place in Soviet biological research than in the West. A number of Soviet research reports have been translated and are available from National Technical Information Service. The Soviet medical research focuses on: clinical effects on the bone (Degen et al., 1971); clinical effects of blood in thrombophlebitis (Edlinskiy et al., 1969; Vyalov, 1971); the progress of infections (Vasil'yev et al., 1971b; Odintsov, 1965); and the progress of tumors (Ukolova and Kvakina, 1971). The biochemical investigations have their focus on enzyme activity (Kugot et al., 1969; Shishlo et al., 1966; Shishlo, 1971; Yashina, 1972), and there is also interest in the effects of magnetism on micro-organisms (Pavlovich, 1971; Kugot, 1969).

Vyalov et al., 1964; Vyalov, 1971

Vyalov reports clinical findings from a group of industrial workers who are exposed to magnetic fields of 10^3 or 10^4 G (50 Hz and DC) during the working day. Vyalov has observed effects of "neurodynamic disturbance" resulting in changed EEG recordings with prominent shifts in the activity of the central nervous system. The neurological findings include changes in vascular tone, changes in perception by the sensory nerves, and hyperhydrosis (excessive sweating).

Another change, also related to the autonomic nervous system, is hypotension due to sinus bradycardia.

Blood chemistry analysis disclosed a disturbed calcium level, and an increased gamma globulin level. Changes in blood composition involved both the white cells (leukopenia or, if a normal leukocyte count, relative lymphocytosis and monocytosis) and the red cells (lower sedimentation rate).

There was a wide spectrum of subjective complaints ranging from poor appetite, inattention and constipation to neurophysiological changes involving muscle or joint pains and strange sensations in the hands.

In a laboratory experiment using DC magnetic fields of 1000 G for a 15-minute exposure period, the researchers found a decrease in the sodium and chlorine ions and an increase in the total calcium content of the blood. These changes are said to be due to the different velocities these ions achieve under the influence of the magnetic field; calcium, being the heaviest ion, moves slowest. The laboratory findings may be correlated to the clinical changes in cardiac muscle function, muscle tone in the peripheral vessels (decreasing the blood pressure) and a change in pain threshold in hands which were exposed to the maximum field. Apparently this experiment did not utilize a control group.

Mykhaylovskiy et al., 1969; Yashina, 1972; Shishlo et al., 1966; Degen, 1971

Mykhaylovskiy et al. (1969): This experiment investigates human susceptibility to very weak fields (apparently about 0.02 G) at frequencies between 0.10 and 2 Hz, and demonstrates that "at least some people can respond to oscillations of geomagnetic field strength."

Yashina (1972): Experiments on rat livers conducted at 7 KHz with a 300 - 900 G pulsed magnetic field indicated an inhibitory effect on the activity of oxidation reduction enzymes and damage to the mitochondrial membrane.

Shishlo *et al.* (1966): Similarly, the oxidative enzymes of the mouse liver were found to have lower levels of activity in experiments conducted at 4000 - 7000 G.

Degen (1971): Bone fractures in rats treated 15 minutes per day with a 500 G constant magnetic field were observed over a period of 7 months. At the end of the first 24 hours there was no effect on the bone, but there were effects on other tissues resulting in less edema and a greater number of phagocytes. After 7 days the periosteum showed more fibrous material than the controls, but by the end of 120 days the differences had vanished. The conclusion is that magnetic fields may hasten bone healing, but do not change the character of the final repair.

Edlinskiy et al., 1969; Kugot et al., 1969; Odintsov, 1965; Tarakhovskiy et al., 1971

Edlinskiy *et al.* (1969) discovered that daily treatments with a constant magnetic field (200 - 400 G for 15 to 30 minutes) produced a clinical improvement in patients suffering from thrombophlebitis and induced an increment in the iron content of the blood and a decrement in the copper content.

Kugot *et al.* (1969): *E. coli* cultivated in a constant magnetic field of 150 G suffered biochemical changes which were first noticed 48 hours after the exposure began.

Odintsov (1965) exposed mice to a 50 Hz magnetic field at 200 G for 6.5 hours and noted a marked increase in resistance to infection either for 1 exposure or 15 repeated daily exposures This experiment, which was controlled, determined resistance to an induced intraperitoneal infection of *Listeria* at various concentrations of the infecting medium, by noting the dose at which 50% of the mice died (LD_{50}). Although no differences occurred during the first two field exposure days, by the third day distinct differences were noted. Odintsov's data show a much lower LD_{50} (20% of the control value), twice as great a level of infection, and a 40% lower phagocytic activity for the leukocytes which themselves numbered only 76% of the quantity in the control group.

Tarakhovskiy et al. (1971) performed an experiment on rats measuring physiological changes under the influence of a constant magnetic field (140 G) or pulsed magnetic field (130 G). They found changes in the rat oestrus cycle and evidence for dystrophy and necrosis in the AC field and marked changes in blood proteins in the DC fields.

Vasil'yev et al., 1971a

This study reports a similarity between reaction to an AC magnetic field (50 Hz) and the ordinary immunological reaction. Albino rats were exposed to a 200 G magnetic field at 50 Hz continously for up to 14 days, and then sacrificed for examination 6, 12, 24, and 48 hours and 3, 5, 7, and 14 days after the fields were energized. A control group of 48 rats was employed. Although no differences were found between the two groups of animals after 6 hours, after 12 hours microscopic examination of the tissues from the spleen, lymph nodes and thymus showed changes along with an indication of an increase of lymphoblasts in the spleen and thymus ($p < 0.05$) and almost a twofold increase in lymphocytes in the thymus ($p < 0.001$). At 24 hours continuing changes in the spleen were observed, but at 48 hours there was no difference between experimentals and controls. There were histological findings for slides of tissues taken after 3, 5, 7, and 14 days, by which time hyperplasia of the lymphoid tissues and increased numbers of lymphocytes, lymphoblasts and neutrophils ($p < 0.001$) were plainly evident. The authors note that "shifts, analogous in their trend but relatively weaker were also observed in the control group." This may be due to the animals' confinement resulting in stress; however, the magnetic field-exposed animals demonstrated considerably more stress reaction. The authors feel that "most probably, these reactions are based on Selye's stress reactions."

Sudakov and Antimoniy, 1973; Dorfman, 1971; Shishlo, 1971

Sudakov and Antimoniy (1973): This article reviews the Soviet and western literature on the effects of electromagnetic fields on the central nervous system with most of the emphasis on the microwave region. (There are 216 references.) The reports at ELF frequencies involve behavioral changes in rats exposed at 200 G (500 Hz) and a shift in cortex potential upon exposure of a cat to several hundred gauss (50 Hz).

Dorfman (1971): This discussion of the effects of magnetism on biological organisms argues that the "magnetotropism" observed in plants results from the mechanical force due to the field inhomogeneity and is not a truly biological effect.

He demonstrates that a magnetohydrodynamic effect on circulating blood is possible. The calculations (Belousova's) show significant effects above several thousand gauss, depending upon the diameter of the blood vessels so that for the large vessels like the aorta the blood flow can be decreased about 10% at 20,000 G, with lesser effects for smaller vessels. Thus, the magnetohydrodynamic effect on blood circulation is probably not an important factor in most magnetic field experiments.

Because myelinated nerve fibers possess an interrupted insulating sheet where the axon is open to the external medium at the nodes of Ranvier, a current flow occurs in the external electrolyte. The current-carrying region is subjected to a force which, it is calculated, results in deflection (bending) of several microns to several hundred microns for a nerve section 1 or 2 mm long in a field of about 10^3 G. Consequently, an elastic wave propagates along a nerve fiber which is discharging in a magnetic field. Dorfman likens the elastic wave to the motion of a string in a stringed musical instrument and speculates that this elastic response of the nerve fibers -- not previously mentioned by others -- may explain magnetic sensitivities observed in persons or animals such as birds or fish.

Dorfman points out that arguments based on the magnetic properties of macromolecules are erroneous because the energy associated with orientation of paramagnetic atoms or molecules, even in fields of 10^3 to 10^5 G, is very small when compared with the thermal energy. However, for diamagnetic macromolecules the orientational energies may be comparable to thermal energies resulting in changes which can be demonstrated in optical phenomena and also affect chemical kinetics, membrane transport, viscosity, etc. He reports an experiment (Haberditzl, 1966) in which an _in vitro_ reaction involving the enzyme glutamate dehydrogenase precipitates 15 or 20% more slowly when in a non-uniform field of about 200 G, "evidently, because of a sharp decrease in the concentration of diamagnetic macromolecules with a molecular weight of about 10^6 in the interpolar volume." At 7×10^4 G, the enzyme precipitation was retarded by 90%.

Dorfman also considers red blood cells which, in the very smallest capillaries, no longer retain their spherical shape and are more susceptible to a magnetic field.

Shishlo, 1971: This study discusses the influence of magnetic fields on enzymes, respiration, and the metabolism of various organisms.

Pavlovich, 1971: Kogan et al., 1971; Novitskiy et al., 1971; Toroptsev et al., 1971

Pavlovich (1971) reviews experiments on micro-organisms exposed to magnetic fields (AC and DC) in which there is contradictory data indicating both inhibition and promotion of culture growth. Experiments he conducted generally indicate a reduction in the activity of the dehydrogenase enzyme in many micro-organism cultures for both DC and AC fields with exceptional cases where enzyme activity was increased. In general, the AC fields were less effective than the DC fields.

His research also indicates that long-term exposures to magnetic fields influence RNA synthesis quantitatively.

Micro-organisms exposed to both high temperatures and magnetic fields showed different death rates for the magnetic field-exposed micro-organisms than for the control, exposed only to high temperatures.

Kogan _et al_. (1971) also discuss experiments on micro-organisms exposed to constant magnetic fields.

Novitskiy _et al_. (1971) discuss experiments on plants under the influence of constant magnetic fields and Pozolotin (1971) reports experiments on plants and seeds exposed to magnetic fields which indicate that chromosomal aberrations can be effected by the presence of the magnetic field.

Toroptsev _et al_. (1971): A histological examination of guinea pigs exposed intermittently (6.5 hours/day) for a period of 1 or 24 days to a 200 G, 50 Hz magnetic field disclosed microscopic changes in the lungs (evidence of weak hemorrhaging or emphysema) and testicles (edema), and evidence of impaired function in the peripheral circulatory system. A single 6.5 hour exposure resulted in changes in the tubules of the testes and the presence of abnormal spermatozoa. Changes in the ovaries of the exposed guinea pigs were less severe but also pathological. Other areas of the body which were found to be abnormal upon investigation were ganglion cells, kidneys, liver, some lymph nodes, bone marrow, and the eye. Examinations made for 30 days after the last field exposure showed almost no signs of damage except in the lungs.

A single 24-hour exposure at 200 G (50 Hz) showed similar but more widespread evidence of damage to tissues. Exposures repeated daily for 24 days gave similar findings.

The experiments conducted in a DC field produced similar results although "restoration processes developed earlier." Finally, the experiment conducted at 7000 G also produced marked pathological effects.

This report presents several slides of tissues showing these pathological changes.

The author's conclusions are that the AC fields have greater biological effectiveness, the testis is the most sensitive organ, the magnetic field had a genetic effect, and the damage that is caused is essentially repaired after the fields are turned off.

(See also the work of Friedman and Carey in Section 5.3.)

Kholodov, 1971

In this work Kholodov discusses magnetic field exposures in relation to the central nervous system, particularly effects on memory, perception, the phosphene phenomenon, reflex activities, and direct effects on neurons. All the experiments he considers are with DC magnetic fields.

Ukolova and Kvakina, 1971

Anti-tumoral effects of magnetic fields due to direct action on the tumor cells or indirect action on the nervous system are reported in this paper. The authors emphasize that the AC magnetic field effects produced by a rotating permanent magnet occur on free radicals. Inhibition was observed for both transplanted tumors and tumors stimulated to grow initially within the host animal. Several micrographs of tumor tissues undergoing resolution under the influence of the rotating magnetic field are presented. The authors discuss Barnothy's similar experiments and draw contrasts between their own findings and his. They feel that the mechanism is "similar to the mechanism of anti-tumoral stimulation of the hypothalamus" resulting in the development of an "activation reaction," a stress-induced syndrome not quite like Selyes' general adaptation syndrome, resulting in increased hormone functions under stimulus of the pituitary gland.

Kholodov, 1970; 1971

Kholodov, who edited the collection of papers which includes the above reports, also presents an extensive bibliography of 407 entries, including Russian and western research. Kholodov has also written a book, Magnetism in Biology (1970), which discusses the role of magnetism in biological research in an historical perspective appropriate to this textbook-like volume which, unfortunately, lacks an index.

Presman, 1974

Presman is the author of a volume, "Electromagnetic Signals in Living Nature: Facts, Hypotheses, Directions of Research" (1974), which discusses electromagnetic fields over the entire frequency spectrum, from ELF to microwaves, in a general fashion illustrated with data from particular experiments. This book contains a bibliography of 101 references but lacks an index. Presman's ideas about electromagnetic communication on the microscopic and macroscopic levels are noteworthy.

8.9 PROPOSED MECHANISMS FOR BIOLOGICAL EFFECTS BY MAGNETIC FIELDS

There are many theories which suggest causes for magnetic interactions with biological tissues. A mechanism affecting molecular reaction rates by a change in the rotational diffusion constants (akin to Brownian motion) was proposed by Valentinuzzi (1962). Barnothy (1964) suggests that because the thermal energy is so much greater than the direct magnetic field interaction energy, the effects are due to a redistribution of diamagnetism and paramagnetism on the molecular level. Maling et al. (1965) and Little (1964) suggest that there are effects on enzyme reactions which resemble the superconducting mechanism in metals at low temperatures. This idea, originally due to F. London (1937; 1950), has also been exploited by Halpern and Wolf (1971). Clement-Metral (1975) suggests that magnetic anisotrophies within the molecules of the eye pigments lead to magnetic rotation allowing perception. At magnetic fields of about 10^5 G, active transport may be affected by changes in the diffusion constant transverse to the field as suggested by Liboff (1965).

Barnothy (1964) also suggests that magnetic fields may have a direct influence on the hydrogen bonds of the double stranded DNA helix and thus may be mutagenic. Gross (1964) argues that small changes in bond angles, under the influence of magnetic fields, influence chemical reaction rates. Vyalov (1971) suggests that because of their differing masses, different ions flow at different rates in a magnetic field leading to changed ionic concentrations. Dorfman (1971) describes magnetohydrodynamic effects on circulation in the large arteries of the body and direct effects on growing cells due to forces developed by a non-uniform magnetic field.

In consideration of the possible effects of a static magnetic field, Abashin and Yevtushenko (1975) present a critical discussion of these and other theories and conclude that there is no case for ferromagnetic effects because of the low concentration of atoms with ferromagnetic properties. Diamagnetic effects are discounted because calculation of the force acting on a macromolecule in an inhomogeneous field indicates no possibility of any effects. Similarly, they doubt the argument that homogeneous fields may affect macromolecular rotation and thereby alter chemical rates. Finally, paramagnetic effects are considered and dismissed on the basis of theoretical and experimental evidence. The authors' final statement, however, indicates that their conclusions "cannot be considered indisputable since our analyses did not allow for the specific of the biological processes..."

8.10 REGULATIONS ON ELECTRIC FIELD EXPOSURES IN THE SOVIET UNION

The Soviet Union has developed rules and regulations for protection of persons exposed to ELF electric fields in industrial situations, "Rules and Regulations on Labor Protection at 400, 500 and 750 kV AC Substations and Overhead Lines of Industrial Frequency (in the U.S.S.R.)"(IEEE, 1975). Laborers working in substations or on transmission lines operating at 400, 500 or 750 kV or otherwise exposed to transmission lines operating at those voltages comprise the only group of persons covered by these rules and regulations. The rules are promulgated because "the electric field...has an unfavorable effect on the human organism [and] besides this at the moment the person touches metallic parts of the equipment, electrical discharges occur." The unfavorable effects of electric fields are limited to regions where the electric field exceeds 5 kV/m so that no restrictions apply at lower field strengths.

The Soviet standards are based upon the assumption that the direct field effect depends both on the field intensity and the duration of exposure so that exposures are regulated with respect to both parameters. No exposure to discharges is permitted. The standards, presented in Table 8-3, show that no exposures over 25 kV/m are permitted.

Where it is not possible to follow the regulations without shielding from the fields, screening or conductive clothing are used to reduce the fields to within the regulated minimums. Particular care is taken to describe the techniques of eliminating discharges by careful grounding of metallic devices, especially vehicles, etc.

In view of these standards there are subsequent regulations on the number of minutes a person may work underneath transmission lines at various distances from the tower. For example, on the 750 kV transmission line, workers are allowed 180 minutes exposure within 30 meters of suspension towers, but no more than 10 minutes in the intermediate span of the line or near anchor and corner towers. In addition, workers in lift buckets or working on the tower must use protective clothing or screens.

The Soviets have not made regulations on magnetic fields, for "the detrimental action of magnetic fields appears at field strengths of 150 - 200 A/m (1.9 - 2.5 G) that can take place only at a distance of 1 - 3 meters from conductors." (Lyskov _et al_., 1975).

These authors state that although standards for the local population and agricultural workers are at present non-existent, they are being developed. Discussing the design of new, higher voltage lines (750 - 1150 kV), they state that these lines "must not be any closer than 300 - 500 meters from the future (15-year planning) borders of population sites, 100 meters from the indivi

TABLE 8-3. RULES AND REGULATIONS ON LABOR PROTECTION AT 400, 500 AND 750 kV AC SUBSTATIONS AND OVERHEAD LINES OF INDUSTRIAL FREQUENCY (In the U.S.S.R.). *

	Electric Field Intensity kV/m	Permissible Duration of Personnel Stay in Electric Field During 24 Hours (minutes)	Notes
1.	5	unlimited	Points 2, 3, 4 and 5 of the regulations are valid if: 1) all the remaining time a man is in areas where electric field intensity is less than or equal to 5 kV/m 2) a possibility of electrical discharge influence is eliminated
2.	10	180	
3.	15	90	
4.	20	10	
5.	25	5	

* IEEE, 1975

TABLE 8-4. MAGNETIC FIELD EXPOSURE STANDARDS*

a) Magnetic Field Safety Standards at Stanford Linear Accelerator

	Extended Period (hours)	Short Periods (minutes)
Whole body or head	200 G (DC)	2000 G (DC)
Arms and hands	2000 G	20000 G

b) Magnetic Field Safety Standards Recommended by A. M. Vyalov

	Field	Field Gradient (Inhomogeneous Field)
Whole body	300 G (DC)	5 - 20 G/cm (DC)
Hands	700 G	10 - 20 G/cm

*Kaufman and Michaelson, 1974

lived-in dwelling, 40 - 60 meters from the old and new non-habitable construction." Other features of the design of these new high voltage lines, e.g., clearances over roads and in difficult terrain, are being designed to the same maximum field levels, 10 - 14 kV/m, used in the 500 kV transmission line network; they cite "favorable operating experience in over 150,000 km-years" with that system.

8.11 REGULATION OF MAGNETIC FIELD EXPOSURES

In the United States there are no governmental standards pertaining to exposures to magnetic fields. At the Stanford Linear Accelerator Center, the laboratory directors have developed a set of regulations limiting exposures to the whole body, head, arms or hands (see Table 8-4a).

In the Soviet Union a similar set of standards has been developed for relatively strong magnetic fields. These regulations specify the maximum field strengths as well as the maximum field gradients of inhomogeneous fields to which the whole body or hands may be exposed (Table 8-4b).

8.12 REFERENCES

Abashin, V.M. and Evtushenko, G.I. 1975. On the problem of the effect of constant magnetic field on biological systems (in Russian). Biofizika 20: 276.

Audus, L.H. and Whish, J.C. 1964. "Magnetotropism." IN: "Biological Effects of Magnetic Fields," Vol. 1. M. Barnothy, ed. Plenum Press, New York.

Azarnia, R., Larson, W.J. and Loewenstein, W.R. 1974. The membrane junctions in communicating and non-communicating cells, their hybrids, and segregants. Proc. Nat. Acad. Sci. (USA) 71: 880.

Barlow, H.B., Kohn, H.E., and Walsh, E.G. 1947. Visual sensations aroused by magnetic fields. J. Physiol. 148: 372.

Barnothy, J.M. 1964. "Development of Young Mice." IN: "Biological Effects of Magnetic Fields," Vol. 1. M. Barnothy, ed. Plenum Press, New York, p. 93.

Barnothy, M.F. 1964. "Biological Effects of Magnetic Fields." Volume 1. Plenum Press, New York.

Barnothy, M.F. 1969. "Biological Effects of Magnetic Fields." Volume 2. Plenum Press, New York.

Barnothy, M.F. and Sumegi, I. 1969. "Effects of the Magnetic Field on Internal Organs and and Endocrine System of Mice." IN: "Biological Effects of Magnetic Fields," Vol. 2. M. Barnothy, ed. Plenum Press, New York, p. 103.

Barnwell, F.H. and Brown, F.A. 1964. "Response of Planarians and Snails." IN: "Biological Effects of Magnetic Fields." Vol. 1, M. Barnothy, ed. Plenum Press, New York, p. 263.

Beal, J.B. 1974. "Electrostatic Fields, Electromagnetic Fields, and Ions." IN: "Biologic and Clinical Effects of Low-Frequency Magnetic and Electric Fields," Llaurado et al., eds. Charles C. Thomas, Springfield, Ill.

Becker, R.O. 1963. Biological effects of magnetic fields: a survey. Med. Electron. Biol. Eng. 1: 293.

Beischer, D.E. 1964. "Survival of Animals in Magnetic Fields of 140,000 Oe." IN: "Biological Effects of Magnetic Fields, Vol. 1, M Barnothy, ed. Plenum Press, New York, p. 201.

Beischer, D.E. 1965. "Biomagnetics." Ann. N.Y. Acad. Sci. 134:454.

Busby, K., Driscoll, D. and Washbon, W.E. 1974. "A Field Survey of Farmer Experience with 765 kV Transmission Lines." Agricultural Resources Commission, State of New York, State Campus, Albany, N.Y

Clement-Metral, J.D. 1975. Direct observation of the rotation in a constant magnetic field of highly organized lamellar structures. FEBS Letters 50: 257.

Cohen, D. 1973. Ferromagnetic contamination in the lungs and other organs of the human body. Science 180: 745.

Cohen, D. 1975a. Measurements of the magnetic fields produced by the human heart, brain and lungs. IEEE Trans. on Magnetics 11: 694.

Cohen, D. 1975b. Magnetic fields of the human body. Physics Today 28: 34.

Cook, E.S. and Smith, M.J. "Increase of Trypsin Activity." IN: "Biological Effects of Magnetic Fields," Vol. 1. M. Barnothy, ed. Plenum Press, New York, p. 246.

Cupcea, S., Deleanu, M. and Frits, T. 1959. Experimentelle Untersuchungen uber den Einfluss der Luftionisation auf Pathologische Veranderungen der Magenschleimhaut. Act. Biol. et Med. Germ. 3: 407.

Degen, I.L. et al. 1971. "Consolidation of Bone Fragments in a Constant Magnetic Field." Ortopediya, Travmatologiya I Protezirovaniye. Available from NTIS as JPRS 62864 (1974).

Dewey, M.M. and Barr, L. 1962. Intercellular connection between smooth muscle cells: a nexus. Science 137: 670.

DiCaprio, R.A., French, A.S., and Sanders, E.J. 1975. Intercellular connectivity in the eight-cell Xenopus embryo. Biophys. J. 15: 373.

Dorfman, Ya. G. 1971. "Physical Phenomena Occurring in Live Objects under the Effect of Constant Magnetic Fields." IN: "Influence of Magnetic Fields on Biological Objects," Y. Kholodov, ed. Available from NTIS as JPRS 63038 (1974).

D'Souza, L., Reno, V.R., Nutini, L.G., and Cook, E.S. 1969. "The Effects of a Magnetic Field on DNA Synthesis by Ascites Sarcoma 37 Cells." IN: "Biological Effects of Magnetic Fields," Vol. 2. M. Barnothy, ed. Plenum Press, New York, p. 53.

Dunlop, D.W. and Schmidt, B.L. 1969. "Sensitivity of Some Plant Materials to Magnetic Fields." IN: "Biological Effects of Magnetic Fields," Vol. 2. M. Barnothy, ed. Plenum Press, New York.

Edlinskiy, I.B. et al. 1969. "Changes in Some Blood Trace Elements in Patients and Acute Thrombophlebitis of the Lower Extremities under the Influence of a Constant Magnetic Field." Voprosy Kurortologii, Fizioterapii I Lechebnoy Fizicheskoy Kul'tury. Available from NTIS as JPRS 62865 (1974).

Fraunfelder, F.T. and Hanna, C. 1972. Electric cataracts I. Sequential changes, unusual and prognostic findings. Arch. Ophthalmol. 87: 179.

Garfield, R.E. and Daniel, E.E. 1974. The structural basis of electrical coupling (cell-to-cell contacts) in rat myometrium. Gynecol. Invest. 5: 284.

Gerencser, V.F., Barnothy, M.F. and Barnothy, J.M. 1964. "Inhibition of Bacterial Growth in Fields of High Paramagnetic Strength." IN: "Biological Effects of Magnetic Fields," Vol. 1. Plenum Press, New York, p. 229.

Gross, L. 1964. "Bibliography of the Biological Effects of Static Magnetic Fields." IN: "Biological Effects of Magnetic Fields," Vol. 1. M. Barnothy, ed. Plenum Press, New York, p. 297.

Gross, W. and Smith, L.W. 1964. "Wound Healing and Tissue Regeneration." IN: "Biological Effects of Magnetic Fields," Vol. 1, M. Barnothy, ed. Plenum Press, New York, p. 140.

Haberditzl, W. 1967. Enzyme activity in high magnetic fields. Nature 213: 72.

Halpern, E.H. and Wolf, A.A. 1971. "Speculations of Superconductivity in Biological and Organic Systems." IN: Advances in Cryogenic Engineering," Vol. 17. K. Timmerhaus, ed. Plenum Press, New York.

Hanna, C. and Fraunfelder, F.T. 1975. "Electrical Cataract." IN: "Cataract and Abnormalities in the Lens." J. Bellows, ed. Grune and Stratton, Inc., New York, p. 224.

Hauf, R. 1974. "Wirkung von 50-Hz-Wechselfeldern auf den Menschen (Effects of 50 Hz alternating fields on man). Sonderdruck aus Elektrotechnische Zeitung-b 26: 381.

Johansson, R. et al. 1973. Is there a connection between the electricity in the atmosphere and the function of man? Part 3, 50 Hz field variations. Forvarets Forskningsanstalt Advelning 2: FOA Report C 2627-H5.

Kaufman, G.E. and Michaelson, S.M. 1974. "Critical Review of the Biological Effects of Low-Frequency Magnetic and Electric Fields." J. Llaurado et al., eds. Charles C. Thomas, Springfield, Ill., p. 49.

Kholodov, Y.A. 1964. "Effects on the Central Nervous System." IN: "Biological Effects of Magnetic Fields," Vol. 1, M. Barnothy, ed. Plenum Press, New York, p. 196.

Kholodov, Y.A. 1970. "Magnetism v Biologii." Izd-vo Nauka, Moscow. Translation of Russian language book available from NTIS as "Magnetism in Biology," JPRS 60737 (1973).

Kholodov, Y.A. 1971. "Vliyaniye Magnitnykh Poley Na Biologicheskiye Ob'yekty," Moscow. Available from NTIS as "Influence of Magnetic Fields on Biological Objects," JPRS 63038 (1974).

Knoll, M., Eichmeier, J., and Schon, R.W. 1964. Properties, measurement, and bioclimatic action of "small" multimolecular atmospheric ions. Adv. Electr. Electron. Phys. 19: 177.

Kogan, A.B. et al. 1974. "The Mechanism of Biological Effects of a Constant Magnetic Field." IN: "Influence of Magnetic Fields on Biological Objects," available from NTIS as JPRS 63038 (1974).

Krueger, A.P., Andriese, P.C., and Kotaka, S. 1968. Small air ions: their effect on blood levels of serotonin in terms of modern physical theory. Int. J. Biometeor. 12: 225.

Krueger, A.P., Kotaka, S., and Andriese, P.C. 1966. "Studies on the Biological Effects of Gaseous Ions: a Review." Special monograph series, Vol. 1. Biometerol. Res. Centre, Leiden, 1.

Kugot, A.S. et al. 1969. "The Effect of a Constant Magnetic Field on E. coli Enzyme Activity.""Izvestiya Sibirskogo Otdeleniya Akademii Nauk SSR." Available from NTIS as "Translations on Biological Effects of Magnetic Fields," JPRS 62865 (1974).

Liboff, R.L. 1965. A biomagnetic hypothesis. Biophys. J. 5: 845.

Little, W.A. 1964. Possibility of synthesizing an organic superconductor. Phys. Rev. 134: A1416.

Loewenstein, W.R. 1966. Permeability of membrane junctions. Ann. N.Y. Acad. Sci. 137 (Art. 2) 403.

Lowenstein, W.R. and Kanno, Y. 1964. Studies on an epithelial (gland) cell junction, I. Modifications of surface membrane permeability. J. Cell Biol. 22: 565.

London, F. 1937. Theorie quantique des courants interatomique dans les combinasons aromatiques. J. Phys. Radium 8: 397-409.

London, F. 1950. "Superfluids," Vol. 1. Dover Publications, Inc., New York, p. 8.

Lyskov, Y.I., Emma, Y.S. and Stolyarov, M.D. 1975. "The Factors of Electrical Field That Have an Influence on a Human." IN: "Three Russian Papers on EHV/UHV Transmission Line and Substation Design." Uhl, Hall and Rich, Div. of Charles T. Main, Inc., Boston, Mass.

Maling, J.E., Weissbluth, M. and Jacobs, E.E. 1965. Enzyme substrate reactions in high magnetic fields. Biophys. J. 5: 767.

Maret, G. et al. 1975. Orientation of nucleic acids in high magnetic fields. Phys. Rev. Letters 35: 397.

McCormick, E. et al. 1971. "Pilot Ecological Field Surveys." Tech. Memo. No. 1, ITRI Project E 6159, IIT Research Institute. Available from NTIS as AD 748334.†

Mericle R.P. et al. 1964. "Plant Growth Responses." IN: "Biological Effects of Magnetic Fields," Vol. 1. M. Barnothy, ed. Plenum Press, New York.

Miller, M.W. 1974. "Effect of Extremely Low Frequency Electric and Magnetic Fields on Roots of Vicia faba." Final Report for Project Sanguine, June, 1974. Dept. of Radiation Biology and Biophysics, Univ. of Rochester, Rochester, N.Y.†

Mulay, I.L. and Mulay, L.N. 1964. "Effect on Drosophila melangaster and S-37 Tumor Cells: Postulates for Magnetic Field Interactions." IN: "Biological Effect of Magnetic Fields." M. Barnothy, ed., Plenum Press, New York, p. 146.

Murr, L.E. 1964. Mechanism of plant-cell damage in an electrostatic field. Nature 201: 1305.

Murr, L.E. 1965. Biophysics of plant growth in an electrostatic field. Nature 206: 467.

Mykhaylovskiy, V.M. et al. 1969. "Human Susceptibility to Weak Magnetic Fields." IN: "Geologiya Geofizika Khimya I Biologiya." Available from NTIS as "Translations on Biological Effects of Magnetic Fields," JPRS 62865 (1974).

Neurath, P.S. 1964. "Simple Theoretical Models for Magnetic Interactions with Biological Units." IN: "Biological Effects of Magnetic Fields," Vol. 1. M. Barnothy, ed. Plenum Press, New York, p. 25.

Novitskiy, Y.I. et al. 1971. "The Influence of Constant Magnetic Fields on the Growth of Plants." IN: "Influence of Magnetic Fields on Biological Objects." Y. Kholodov, ed. Available from NTIS as JPRS 63038 (1974).

Odintsov, Y.N. 1965. "The Effect of a Magnetic Field on the Natural Resistance of White Mice to Listeria Infection." "Voprosy Epidemiologii Mikrobiologii l Immunologii." Available from NTIS as JPRS 62865 (1974).

Pammenter, N.W., Adamson, J.H. and Berjak, P. 1974. Viability of stored seed: extension by cathodic protection. Science 186:1123

Pavlovich, S.A. 1971. "The Influence of Magnetic Fields on Microorganisms." IN: "Influence of Magnetic Fields on Biological Objects." Y. Kholodov, ed. Available from NTIS as JPRS 63038 (1974).

Pozolotin, A.A. 1971. "The Influence of Magnetic Fields on Radiation-Induced Chromosomal Aberrations in Plants." IN:

"Influence of Magnetic Fields on Biological Objects." Y. Kholodov, ed. Available from NTIS as JPRS 63038 (1974).

Presman, A.S. 1974. "Electromagnetic Signals in Living Nature: Facts, Hypotheses, Directions of Research." Translation from Russian book available from NTIS as JPRS 64228 (1975).

Shishlo, M.A. 1971. "Influence of Magnetic Fields on Enzymes, Tissue Respiration and Some Aspects of Metabolism in an Intact Organism." IN: "Influence of Magnetic Fields on Biological Objects," Y. Kholodov, ed. Available from NTIS as JPRS 36038 (1974).

Shishlo, M.A. *et al*. 1966. "The Effect of Exposure of the Intact Organism to a Constant Magnetic Field on the Activity of Oxidative Enzymes in the Livers of Mice." IN: "Patologicheskaya Fiziologiya." Available from NTIS as "Translations on Biological Effects of Magnetic Fields," JPRS 62865 (1974).

Sidaway, G.H. 1966. Influence of electrostatic fields on seed germination. *Nature* 211: 303.

Spanswick, R.M. 1972. Electrical coupling between cells of higher plants: a direct demonstration of intercellular communication. *Planta* (*berl*.) 102: 215.

Strumza, M.V. 1970. Influence sur la santé humaine de la proximité des conducteur d'électricite a haute tension. (Influence of proximity to high tension electrical lines on human health.) *Arch*. *Mal*. *Prof*. (Paris) 31: 269.

Sudakov, K.V. and Antimoniy, G.D. 1973. "Central Mechanism of Action of Electromagnetic Fields." IN: "Uspekhi Fiziologicheskikh Nauk," Russian, No. 2: 101. Available from NTIS as JPRS 60711 (1973).

Tarakhovskiy, M.L. *et al*. 1971. "Effect of Constant and Variable Magnetic Fields on Some Indices of Physiological Functions and Metabolic Processes in White Rats." IN: "Fiziologichyy Zhurnal Akademiyi Nauk Ukrayins'koyi RSR." Available from NTIS as "Translations on Biological Effects of Magnetic Fields," JPRS 62865 (1974).

Toroptsev, I.V. *et al*. 1971. "Pathologoanatomic Characteristics of Changes in Experimental Animals under the Influence of Magnetic Fields." IN: "Influence of Magnetic Fields on Biological Objects." Y. Kholodov, ed. Available from NTIS as JPRS 63038 (1974).

Ukolova, M.A. and Kvakina, Y.B. 1971. "Effect of Magnetic Fields on Experimental Tumors (Direct and through the Nervous System). IN: "Influence of Magnetic Fields on Biological Objects," Y. Kholodov, ed. Available from NTIS as JPRS 63038 (1974).

Valentinuzzi, M. 1972. Theory of magnetophosphenes. Am. J. Med. Electronics 1: 112.

Vasil'yev, N.V., Boginich, L.F. and Shilyayeva, G.G. 1971a. Elements of similarity in the reaction of the lymphoid tissue to antigenic and nonantigenic (alternating magnetic field) stimulation. Moscow Arkhiv Patologii 9: 47. Available in translation from Russian from NTIS as JPRS 62510 (1974).

Vasil'yev, N.V. et al. 1971b. "Magnetic Fields, Infection and Immunity." IN: "Influence of Magnetic Fields on Biological Objects." Y. Kholodov, ed. Available from NTIS as JPRS 63038 (1974).

Vyalov, A.M. et al. 1964. "On the Problems of the Effect of Constant and Variable Magnetic Fields on the Human Organism." IN: "Voprosy Profpatologii" (Problems of Occupational Pathology). A. Shitskova, ed. Ministry of Health, U.S.S.R., Moscow. Available in translation from Russian from NTIS as JPRS 36192, (1966).

Vyalov, A.M. 1971. "Clinico-Hygienic and Experimental Data on the Effects of Magnetic Fields under Industrial Conditions." IN: "Influence of Magnetic Fields on Biological Objects." Y. Kholodov, ed. Available from NTIS as JPRS 63038 (1974).

Ware, B.J. 1975. "Effect of 765 kV Transmission Lines on Animal Grazing Habits." American Electric Power Service Corporation, December, 1974. (Rev., 1975).

Wiener, J., Spiro, D., and Loewenstein, W.R. 1964. Studies on an epithelial (gland) cell junction. II. Surface structure. J. Cell Biol. 22: 587.

Wiley, R.H., Cooke, S.L., Crawford, T.H., Fairless, B.J., Liu, H., and Weber, E.C. 1964. "Magnet Reactivation of Partially Inhibited Trypsin." IN: "Biological Effects of Magnetic Fields," Vol. 1. M. Barnothy, ed. Plenum Press, New York, p. 255

Worden, J.L. 1961. Proliferation of mammalian cells in ion-controlled environments. J. Natl. Cancer Inst. 26: 801.

Yashina, L.H. 1972. "The Effect of a Low-Frequency Pulsed Magnetic Field on the Activity of Oxidation-Radiation Enzymes in the Liver of Albino Rats (A Historical Study). IN: "Gigyena Truda I Professional'nye Zabolvaniya." Available from NTIS as "Translations on Biological Effects of Magnetic Fields," JPRS 628 (1974).

Young, W. 1969. "Magnetic Field and in Situ Acetylcholinesterase in the Vagal Heart System." IN: "Biological Effects of Magnetic Fields," Vol. 2. M. Barnothy, ed. Plenum Press, New York, p. 79

Young, W. and Gofman, J.W. 1965. "Magnetic Fields, Vagal Inhibition, and Acetylcholinesterase Activity." Lawrence Radiation Laboratory, Univ. of California--Livermore.

Chapter 9

TABULAR SUMMARY OF PRINCIPAL INVESTIGATIONS

9.1 TABLE 9-1. ELECTRIC FIELD EFFECTS MEASURED BY TOTAL CURRENT FLOW

	Current (A)	Reference	Section
1. Total current flow at 60 Hz calculated from a model of the body as a prolate spheroid located far from ground in a vertical, uniform 10 kV/m electric field			
a) Man	a) 6.5×10^{-5}	Barnes et al., 1967	4.5
b) Rat	b) 4.5×10^{-7}		
c) Mouse	c) 1.4×10^{-7}		
2. 60 Hz current in a man standing on the ground in a vertical 10 kV/m electric field	2.0×10^{-4}	Deno and Zaffanella, 1975	4.5
3. a) Maximum 50 Hz current flow among employees of Soviet 500 kV substation	2.3×10^{-4}	Asanova and Rakov, 1966	4.5
b) Average current flow in substation workers	$1.15 - 1.25 \times 10^{-4}$		
4. 60 Hz current flow in linemen during live-line maintenance of transmission lines operating at 138 and 345 kV. Various amounts of shielding are employed	$0.7 - 4 \times 10^{-4}$	Kouwenhoven et al., 1966	5.2
5. DC current flow which affects RNA synthesis in frog erythrocytes	$2.5 - 6.0 \times 10^{-10}$	Harrington and Becker, 1973	5.7

TABLE 9-1. ELECTRIC FIELD EFFECTS MEASURED BY TOTAL CURRENT FLOW.(Cont'd)

	Current (A)	Reference	Section
6. DC current affecting			
a) Bone healing in a human being	a) 3.9×10^{-6}	Lavine et al., 1972	5.10
b) Bone growth and healing in rabbit femur	b) 4×10^{-6}	Rinaldi, 1974	5.10
c) Bone morphology and metabolism in humans	c) 3.5×10^{-6}	Lavine, 1974	5.10
d) Bone growth and strontium uptake	d) $2.5 - 40 \times 10^{-6}$	Klems, 1975	5.10
7. DC current used in regeneration studies			
a) Limb regeneration in frogs	a) 1.0×10^{-7}	Smith, 1974	7.24
b) Limb regeneration in rats	b) $10^{-10} - 10^{-8}$	Becker and Spadaro, 1972	7.24
c) Causing pathological changes in rat leg	c) $5 - 15 \times 10^{-9}$	Becker, 1972c	7.24

9.2 TABLE 9-2. ELECTRIC FIELD EFFECTS MEASURED BY CURRENT DENSITY

	Current Density (A/m^2)	Reference	Section
1. Fish sensitivity*			
American eels	1.7×10^{-6} to 1.7×10^{-5}	Rommel and McCleave, 1973a	7.14
American salmon	1.5×10^{-7} to 1.5×10^{-6}		
2. Growth and survivability of neural tissue	1.5×10^{-6} to 20×10^{-3}	Sisken and Smith, 1975	7.24
3. Average current density for whole body exposure in a 10 kV/m electric field (remote from ground or other objects)			
Man	520×10^{-6}	Barnes et al., 1967	4.5, Table 4.2
Rat	75×10^{-6}		
Mouse	76×10^{-6}		
4. Man standing on ground in a 10 kV/m electric field	1.6×10^{-3}	**	4.5
5. Substation workers in Soviet study (average value)	1×10^{-3}	**	5.2

* Current density in water of varying salinity

** These values are scaled from the current densities in the prolate spheroidal model (Barnes et al., 1967) according to the relative total currents (See Table 4-2)

† Current density in cortex calculated from electric field and an assumed cortex resistivity of 300 ohm-cm

†† Current density in sea water

TABLE 9-2. ELECTRIC FIELD EFFECTS MEASURED BY CURRENT DENSITY (Cont'd)

	Current Density (A/m^2)	Reference	Section
6. American linemen in AEP study	0.5 to 2.0×10^{-3}	**	5.2
7. Mice in 160 kV/m field	1.2×10^{-3}	**	7.10
8. Electrical activity of algae	60×10^{-3}	Jaffe, 1966	7.18
9. Alteration of Ca^{++} and GABA efflux in brain tissue using electrodes†	1.7	Kaczmarek and Adey, 1974	5.6
10. Effects on metabolism in mitochondria	18	Riesen et al., 1972	5.4
11. Behavioral effects in marine animals††	10 - 650	Straub et al., 1972	7.15
Death or harm to some fish	35 - 80	Straub et al., 1972	7.15

† Current density in cortex calculated from electric field and an assumed cortex resistivity of 300 ohm-cm

†† Current density in sea water

9.3 TABLE 9-3. ELECTRIC FIELD EFFECTS MEASURED BY ELECTRIC FIELD STRENGTH

	Effect Observed? + = yes - = no	Electric Field (V/m)	Reference	Section
1. Typical field values				
Near common household appliances		10-100	D.A. Miller, 1974	3
Near transmission line right of way (typical)		0.5-2 $\times 10^3$	Deno and Zaffanella, 1975	3
Typical maximum ground level fields near transmission lines (345 kV-765 kV)		4-15 $\times 10^3$	Deno and Zaffanella, 1975	3
Spark discharges occur		> 2-5 $\times 10^3$	Zaffanella, personal communication	
2. Studies in humans				
Project Sanguine workers near antenna at Wisconsin Test Facility (45-75 Hz)	-	∿ 5 $\times 10^{-2}$	Krumpe and Tockman, 1972	5.2
Students in laboratory study of pulse, respiration, EEG and blood count	+,-	1-20 $\times 10^3$	Hauf, 1974	5.2
Soviet substation workers (average values for maintenance personnel)	+	7.8-10 $\times 10^3$	Asanova and Rakov, 1966	5.2
Soviet substation workers (maximum values for maintenance personnel)		14.5 $\times 10^3$	Asanova and Rakov, 1966	5.2
Psychological tests	-	20 $\times 10^3$	Johansson et al.,1973	6.8
American linemen	-	>25 $\times 10^3$	Est. from Barnes et al., 1967, Table II	5.2

TABLE 9-3. ELECTRIC FIELD EFFECTS MEASURED BY ELECTRIC FIELD STRENGTH (Cont'd. - 2)

	Effect Observed? + = yes - = no	Electric Field (V/m)	Reference	Section
3. Perception in humans				
Reaction times at 3-10 Hz	+	1-5	Konig, 1971	6.1
Reaction times at 2, 6 and 12 Hz	+	4	Hamer, 1968	6.1
Forced choice test - 40, 400, 4000 Hz	-	650	Milburn, 1971	6.1
Spark discharges felt	+	3×10^3	Takagi and Muto, 1971	6.7
Spark discharges painful	+	6×10^3	"	
4. Physiological effects in humans				
Accelerated biological rhythm (10 Hz square wave)	+	2.5	Wever, 1967, 1968, 1974	6.9
Blood pressure and galvanic skin changes due to discharges	+	2.5×10^3	Takagi and Muto, 1971	6.7
Nervous system disorders in substation workers in USSR (50 Hz)	+	$8\text{-}14.5 \times 10^3$	Asanova and Rakov, 1966	6.5
Neurological effects related to spark discharges	+	10×10^3	Krivova, 1968, Krivova et al., 1975a	6.6
5. Laboratory studies of whole animals				
Pathological changes in two generations of rats (45, 75 Hz, also 1 or 2 G magnetic field)	-	10, 20	Coate et al., 1970	7.19
Fertility and neonatal survival in rats (45, 75 Hz, also 1 or 2 G magnetic fields)	-,+	10, 20	Coate et al., 1970	7.19

TABLE 9-3. ELECTRIC FIELD EFFECTS MEASURED BY ELECTRIC FIELD STRENGTH (Cont'd. - 3)

	Effect Observed? + = yes - = no	Electric Field (V/m)	Reference	Section
Survival and growth in 3 generations of mice (60 Hz)	+	$10\text{-}15 \times 10^3$	Marino et al., 1976b	5.3
Blood proteins, glaucoma - in rats (DC)	+	$0.6\text{-}19.7 \times 10^3$	Marino et al., 1974a	5.3
Blood proteins, steroids and weight - in rats @ 60 Hz	+	15×10^3	Marino et al., 1976c	5.3
6. Metabolic effects				
Respiration in mitochondria @ 60 Hz	-	6.3	Riesen et al., 1971	5.4
Respiration in mitochondria @ 60 Hz	+	155	Riesen et al., 1971	5.4
DNA Synthesis, and collagen production in fibroblasts (DC)	+	10^5	Bassett and Hermann, 1968	5.7
Increased immunity in mice	+	2.5×10^6	Fischer, 1973	5.8
7. Effects on bone				
Increased mitosis in healing bone (DC)	+	-	Rinaldi et al., 1974	5.10
Bone healing and bone strength (@ 1 and 65 Hz)	+	10^{-2} to 5	Bassett et al., 1974[a], 1974[b]	5.10
Bone growth and morphology (@ 5 Hz)	+	$2.7\text{-}8.1 \times 10^4$	Norton, 1974	5.10
Strength of bone (DC, 3, 30 Hz)	+	$\sim 10^4$ V/m	McElhaney et al., 1968	5.10
Tumors in bone (DC, 3, 30 Hz)	+	$\sim 10^4$ V/m	"	5.10

TABLE 9-3. ELECTRIC FIELD EFFECTS MEASURED BY ELECTRIC FIELD STRENGTH (Cont'd. - 4)

	Effect Observed? + = yes - = no	Electric Field (V/m)	Reference	Section
8. Effects on animal behavior				
Electroreception in fish	+	10^{-6}	Bullock, 1973	7.14
Threshold for effects on conditioned behavior in fish	+	$5\text{-}50 \times 10^{-6}$	Rommel and McCleave, 1973 a	7.14
Perception and aversion by fish, ducklings, turtles, rats and fruit flies (at 45 and 75 Hz, also with a 1 or 2 G magnetic field)	+,-	10, 20	Coate et al., 1970	7.23
Prenatal influence on rat avoidance behavior (at 45, 75 Hz, also with a 1 or 2 G magnetic field)	-	20	Coate et al., 1970	7.23
Interresponse time in monkeys (7, 10, 45, 60, 75 Hz)	+,-	1-100	Gavalas et al., 1970, Gavalas-Medici and Magdaleno, 1975	7.1
Hamsters show avoidance behavior at 10 Hz to 10 MHz	+	100-900	Schua, 1953, 1954	7.23
Planarian motion (DC field in air)	+	200	Brown, 1962a	7.23
Locomotion in snails (DC)	+ +	200 $1.5\text{-}5 \times 10^{3}$	Webb et al., 1959, 1961	7.23
Biorhythms and activity of fruit flies	+	$300, 1\text{-}7 \times 10^{3}$	Picton, 1966	7.23
Activity in mice Increased mortality in irradiated mice	+ ?	800-1500 10^{3}	Moos, 1964 Moos, 1964	7.11 7.11
Activity in Drosophila (DC)	+	$2\text{-}6 \times 10^{3}$	Edwards, 1960	7.23

TABLE 9-3. ELECTRIC FIELD EFFECTS MEASURED BY ELECTRIC FIELD STRENGTH (Cont'd. - 5)

	Effect Observed? + = yes - = no	Electric Field (V/m)	Reference	Section
9. Physiological effects in animals				
Blood and urine chemistry in dogs (at 45, 75 Hz; also a 1-2 G magnetic field and a 0.5 mA current)	-	10, 20	Coate et al., 1970	7.20
Blood pressure and thermal regulation in dogs (at 45, 75 Hz; also a 1-2 G magnetic field and a 0.5 mA current)	+?	10, 20	Coate et al., 1970	7.20
ECG changes in rats (DC)	+	10^4	Lott and McCain, 1973	7.6
ECG changes in mice and guinea pigs (50 Hz)	+	10^5	Blanchi et al., 1973b	7.5
EEG changes in guinea pigs (50 Hz)	+	10^5	Blanchi et al., 1973b	7.5
10. Effects on nerves or neural tissue				
Alteration of neuronal firing rate (field across neuron)	+	< 1	Terzuolo and Bullock, 1956	7.8
GABA efflux in cat brain (field in brain tissue)	+	5	Kaczmarek and Adey, 1974	5.6
Ca^{++} efflux in chick and cat brain (1, 6, 16, 32, 75 Hz) (field in air)	+,-	5-56	Bawin and Adey, 1975	5.6
Direct stimulation of nerve in air	+	5.5×10^4	Gengerelli and Holter, 1941	7.8
Direct stimulation of nerve in saline solution	-	5.5×10^4	Gengerelli and Holter, 1941	7.8

TABLE 9-3. ELECTRIC FIELD EFFECTS MEASURED BY ELECTRIC FIELD STRENGTH (Cont'd. - 6)

	Effect Observed? + = yes - = no	Electric Field (V/m)	Reference	Section
Direct stimulation of rat nerve *in vivo*	-	2.5×10^{5}	Gengerelli and Holter, 1941	7.8
Calculated field value for direct effect on membrane potential		$> 10^{7}$	Schwan, 1972	4.10
11. Effects on reproductive mechanisms				
Mitotic delay in slime mold (at 60 and 75 Hz combined with a 1 G magnetic field)	+	7×10^{-1}	Marron et al., 1975	7.9
Mitotic delay in slime mold (at 75 Hz, and 0.4 G magnetic field)	-	1.5×10^{-1}	Marron et al., 1975	7.9
Mutagenesis in fruit flies (45, 75 Hz, also 2 G magnetic field)	+	20	Coate et al., 1970	7.21
Mutagenesis in fruit flies (45, 60, 75 Hz, also 1 G magnetic field)	-	10	Mittler, 1972	7.21
Mutagenesis in fruit flies (45, 75 Hz, also a 2 G magnetic field)	-	20	Bender, 1972	2.4
Mutagenesis in *E. coli*	-	10, 20	Coate et al., 1970	7.22
Fertility, hatchability, and survivability of chick embryos at 60, 75 Hz	-	1, 3500	Durfee, 1975a	7.17
Chromosomal abnormalities in tumor cells implanted in mice exposed to field (DC)	+	$7\text{-}15 \times 10^{3}$	Marino et al., 1974b	5.3

TABLE 9-3. ELECTRIC FIELD EFFECTS MEASURED BY ELECTRIC FIELD STRENGTH (Cont'd. - 7)

	Effect Observed? + = yes - = no	Electric Field (V/m)	Reference	Section
12. *In vitro* studies				
Electrical properties of embryonic growth of plants (DC)	+	1.3×10^{-1}	Jaffe, 1966; Robinson & Jaffe, 1975; Weisenseel et al., 1975	7.18
Growth inhibition at 75 Hz	+	1	Yates et al., 1975	7.17
Growth inhibition and enhancement at 60 Hz	+	10	Yates et al., 1975	7.17
Plant embryo orientation in 8 Hz electric field	+	-	Novak and Bentrup, 1973	7.18
Amoeba change shape (reversibly)	+	$< 1.5 \times 10^{3}$	Friend et al., 1975	7.25
Amoeba may be harmed or killed	+	$> 1.5 \times 10^{3}$	Friend et al., 1975	8.25
13. Surveys				
Of persons living near high voltage transmission lines in France (50 Hz)	-	$< \sim 7 \times 10^{3}$	Strumza, 1970	8.4
Of farmers, effects on plants and animals (60 Hz)	-	$< 10^{4}$	Busby et al., 1974	8.4
Of farmers with grazing animals (60 Hz)	-	$< 10^{4}$	Ware, 1975	8.4

9.4 TABLE 9-4. MAGNETIC FIELD EFFECTS (BELOW ~ 200 G)

	Effect Observed? + = yes - = no	Magnetic Flux Density (gauss)	Reference	Section
1. Typical Magnetic Fields				
Natural Geomagnetic Field at 60 Hz		10^{-8}	Konig, 1974a	3
Near electrical appliances at 60 Hz		10^{-3} - 25	D.A. Miller, 1974	3
At edge of transmission line right of way (765 kV line carrying 4000 A)		0.25 - 0.5	Deno, 1975	9.2
Maximum under 765 kV transmission line (carrying 4000 A)		0.3 - 0.8	Deno, 1975	9.2
Near (< 2 m) conductor of transmission line		1 - 10	My calculation (ARS)	3
2. Tests on humans				
Change in flicker fusion frequency in "zero" field	+	10^{-3}	Beischer, 1965	6.4
Psychological and physiological changes in "zero" field	-	10^{-3}	" "	6.4
Medical evaluation of workers at Project Sanguine test antenna (45 - 75 Hz)	-	0.01 - 0.15	Krumpe and Tockman, 1972	5.2, 6.8
Medical psychomotor and behavioral evaluation of Navy pilots, elevation of serum lipids (45 Hz)	-,+	1	Beischer et al. 1973	5.2
Performance on tests of cognitive ability (45 Hz)	+	1	Gibson and Moroney, 1974	6.3
Detection of field by a forced choice test (40, 400, 4000 Hz)	-,+	.05	Milburn, 1971	6.1

TABLE 9-4. MAGNETIC FIELD EFFECTS (BELOW ~ 200 G) (Continued - 2)

	Effect Observed? + = yes − = no	Magnetic Flux Density (gauss)	Reference	Section
2. Tests on humans (cont'd.)				
Perception using whole body exposures (60 Hz)	+	7 - 10	Schmitt and Tucker, 1973	6.1
Perception using exposures to the head only (60 Hz)	-	15	" "	6.1
Changes in reaction time (0.1 and 0.2 Hz)	+	5 - 11	Friedman et al. 1967	6.1
3. Whole body studies in animals				
Respiration and viability of soil animals at site of Project Sanguine test antenna (45 - 75 Hz)	-	$1\text{-}3 \cdot 10^{-1}$	Greenberg, 1975	5.5
Fertility in rats (45, 75 Hz)	- ?	1,2	Coate et al. 1970	7.19
Pathology in rats (also a 10 or 20 V/m electric field)	-	1,2		
Physiology and blood chemistry in dogs (excepting below) (also a low level electric field and an 0.5 mA current)	-	1,2	Coate et al. 1970	7.20
Temperature regulation and hypertension in dogs (also a low level electric field and an 0.5 mA current)	+ ?	1,2	" "	7.20
Serum triglyceride levels in rhesus monkeys (15 and 45 Hz)	-	9	deLorge, 1974	5.2
Serum triglycerides in mouse liver (45 Hz)	-	1	Beischer and Brehl, 1975	5.2
Steroid changes in rats (50 Hz)	+	200	Udintsev and Moroz, 1974	5.3

TABLE 9-4. MAGNETIC FIELD EFFECTS (BELOW ~200 G) (Continued - 3)

	Effect Observed? + = yes - = no	Magnetic Flux Density (gauss)	Reference	Section
3. Whole body studies in animals (Cont'd.)				
Stress effects in monkeys and rabbits (DC)	+ ?	200	Friedman and Carey; 1969,1972	5.3
Neuropathology in monkeys and rabbits (DC)	-	200	"	5.3
4. Behavioral Responses in animals				
Magnetic compass used by homing pigeons	+	0.1 to 0.5	Walcott, 1974	7.14
Compass orientation of gull chicks (45 - 75 Hz) (also with an electric field of 0.2 to 0.3 V/m)	+	0.5	Southern, 1973, 1975	7.14
Conditioned behavior of pigeons and rats (also an electric field of 0.06 to 100 V/m)	-	0.1 to 1.5	Marr et al.,1973	7.16
Suppression of learning in rats due to a prenatal exposure to a rotating magnetic field (0.5 Hz)	+	0.5 - 30	Persinger and Pear, 1972, Persinger et al.,1974	7.13
Prenatal influence on rat avoidance behavior (45 and 75 Hz)	-	1,2	Coate et al. 1970	7.23
Perception and aversion in fish, ducklings, turtles, rats and fruit flies	-,+ ?	1,2	Coate et al. 1970	7.23
Stimulus-response time in monkeys (7 and 45 Hz)	-	3	Grissett and deLorge, 1971	7.4
Activity level, usefulness as a cue and interresponse time test in monkeys (15 Hz)	-	9	deLorge, 1974	7.4

TABLE 9-4. MAGNETIC FIELD EFFECTS (BELOW ~200 G) (Continued - 4)

	Effect Observed? + = yes - = no	Magnetic Flux Density (gauss)	Reference	Section
4. Behavioral Responses in animals (Cont'd.)			deLorge, 1974	7.4
Fixed ratio response in monkeys (45 Hz)	+ ?	9		
Fixed interval response, reaction times, matching tests, performance tasks, activity levels and health in monkeys (10, 45, 60 and 75 Hz; combined exposure with low level electric field)	-	9	deLorge, 1973a, 1973b	7.4
Decreased activity in 2 monkeys (10 Hz) (combined exposure with low level electric field)	+	9	deLorge, 1973a	7.4
5. Effects on Reproductive System, Growth and Development				
Mutagenesis in fruit flies (also a 10 or 20 V/m electric field)	+	2	Coate et al., 1970	7.21
Mutagenesis in fruit flies (also a 10 V/m electric field)	-	1	Mittler, 1972	7.21
Mutagenesis in fruit flies (also a 20 V/m electric field)	-	2	Bender, 1972	2.4
Mitotic delay in slime mold (60 and 75 Hz) (also a 0.7 V/m electric field)	+	1	Marron et al., 1975	7.9
Mitotic delay in slime mold (60 and 75 Hz) (also a 0.15 V/m electric field	-	0.4	"	7.9

9.5 TABLE 9-5. EFFECTS IN STRONG MAGNETIC FIELDS (ABOVE ~ 200 G)

	Effect Observed? + = yes - = no	Magnetic Flux Density (gauss)	Reference	Section
1. Physiological Effects in Humans				
Magnetophosphenes (10 - 100 Hz)	+	100	Barlow, 1947 Valentinuzzi, 1962	8.8
Clinical improvement in thrombophlebitis and associated changes in Fe and Cu in blood (DC)	+	200 - 400	Edlinskiy, 1969	8.8
Effects on Na^+, Cl^- and Ca^{++} in blood (DC)	+	10^3	Vyalov, 1971	8.8
Electroencephelagram changes, effects on central nervous system and blood chemistry (DC, 50 Hz)	+	$10^3 - 10^4$	Vyalov et al. 1964	8.8
2. Effects of Whole Body Exposures on Animal Tissues				
Altered blood proteins in rats (DC)	+	500	Degen et al., 1971	8.8
Stress effects observed in spleen, thymus and lymph tissues, similarity to immunological reaction (50 Hz)	+	200	Vasil'yev et al., 1971a,b	8.8
Improved resistance to infection (50 Hz)	+	200	Odintsov, 1965	8.8
Faster healing of bone in rats (DC)	+	500	Degen et al., 1971	8.8
Tissue pathology in guinea pig (lungs, testicles, ovaries), circulatory system changes (50 Hz and DC)	+	$200, 7 \cdot 10^3$	Kogan et al., 1971	8.8
Effects on wound healing but not on tissue regeneration	+	$3 - 4\ 10^3$	Gross and Smith 1964	8.8

TABLE 9-5. EFFECTS IN STRONG MAGNETIC FIELDS (ABOVE ~200 G) (Continued - 2)

	Effect Observed? + = yes - = no	Magnetic Flux Density (gauss)	Reference	Section
2. Effects of Whole Body Exposures on Animal Tissues (cont'd.)				
Lethal effects in mice, interference with white blood cell formation (inhomogeneous DC field)	+	4.2×10^3	Barnothy, 1964	8.8
Enhanced immune response, tumor rejection and higher activity levels in mice (inhomogeneous DC field)	+	4.3×10^3	Barnothy, 1964	8.8
3. Effects on Biosynthesis and Biochemical Activity				
Increased acetylcholinesterase activity in frog heart	+	4×10^3	Young and Gofman, 1965	8.8
Altered enzyme activity in mouse liver	+	$4 - 7 \times 10^3$	Shishlo et al., 1966	8.8
Lower rate of DNA synthesis and respiration in tumor cells	+	7.5×10^3	D'Souza et al. 1969	8.8
4. Functional and Behavioral Effects				
Alteration of oestrus cycle in rats, dystrophic and necrotic effects on tissue (pulsed DC)	+	125	Tarakhovskiy et al., 1971	8.8
Behavioral changes in rats and shift in cortex potential of cats (50 Hz)	+	200	Sudakhov and Antimoniy, 1973	8.8
EEG changes in rabbits (DC)	+	$0.2 - 1 \times 10^3$	Kholodov et al., 1969	9.2
Conditioned response in rabbits (inhomogeneous DC field)	+	10^3	" "	9.2

TABLE 9-5. EFFECTS IN STRONG MAGNETIC FIELDS (ABOVE ~200 G) (Continued - 3)

	Effect Observed? + = yes - = no	Magnetic Flux Density (gauss)	Reference	Section
4. Functional and Behavioral Effects (Cont'd.)				
Amplitude and frequency changes in monkey EEGs	+	$2 - 9.1 \times 10^4$	Beischer and Knepton, 1966	7.7
Observations of blood flow, altered electro-cardiograms in monkeys	+	$0.4 - 1 \times 10^5$	Beischer, 1969	9.2
5. Effects on Reproduction and Development				
Mutagenesis in Drosophila (DC)	+	$1 - 5.2 \times 10^2$	Tegenkamp, 1969	9.2
Inhibition of development in frog embryos (inhomogeneous DC field)	+	10^4	Neurath, 1969	9.2
Mutagenesis in Drosophila, mice, bacteria (DC)	-	1.4×10^5	Beischer, 1964	8.8
Embryonic development of sea urchin eggs (DC)	+	$1.4 \cdot 10^5$	Beischer, 1964	8.8
6. Physical and Chemical Effects				
"Superconducting" behavior (anomalous dielectric constant) in protein molecules (DC)	+	600 - 850	Ahmed et al., 1975	2.4
In vitro inhibition of enzyme reactions (DC)	+	$2 \cdot 10^2 - 8 \cdot 10^4$	Haberditzl, 1966 Dorfman, 1971	2.4
Orientation of DNA and synthetic polymers (DC)	+	10^5	Maret et al., 1975	8.7

9.6 REFERENCES

References cited in Tables 9-1 through 9-5 are to be found at the end of the chapter of the section indicated after each reference name.

The following references do not appear in the text proper and are given below:

Beischer, D.E. 1969. "Vectorcardiogram and Aortic Blood Flow of Squirrel Monkeys (Saimiri sciureus) in a Strong Superconductive Electromagnet." IN: "Biological Effects of Magnetic Fields," Vol. 2, M. Barnothy, ed. Plenum Press, New York.

Deno, D. 1975. "Common Record Hearings on Health and Safety of 765 kV Transmission Lines." Testimony before Public Service Commission of the State of New York in cases 26529 and 26559.

Kholodov, Y. et al. 1969. "Investigation of the Reactions of Mammalian Brain to Static Magnetic Fields." IN: "Biological Effects of Magnetic Fields," Vol. 2, M. Barnothy, ed. Plenum Press, New York, p. 215.

Neurath, P.S. 1969. "The Effects of High Gradient, High Strength Magnetic Fields on the Early Embryonic Development of Frogs." IN: "Biological Effects of Magnetic Fields," Vol. 2, M. Barnothy, ed. Plenum Press, New York, p. 177.

Tegenkamp, T.R. 1969. "Mutagenic Effects of Magnetic Fields on Drosophila melanogaster." IN: "Biological Effects of Magnetic Fields," Vol. 2, M. Barnothy, ed. Plenum Press, New York.

Chapter 10

GENERAL BIBLIOGRAPHY

The following references were useful in providing background information for the authors, but were not cited specifically in the text. They are included below as general bibliography, since they may be helpful to the reader wishing to pursue further lines of research in this field.

Aceto, H., Tobias, C.A. and Silver, I.L. 1970. Some studies on the biological effects of magnetic fields. IEEE Trans. on Magnetics MAG-6: 368-373.

Altmann, G. 1974. Oxygen and biochemical changes following ELF exposure. IN: "ELF and VLF Electromagnetic Field Effects." M. Persinger, ed. Plenum Press, New York, p. 227.

Amer, N.M. and Tobias, C.A. 1965. Analysis of the combined effect of magnetic fields, temperature and radiation on development. Biophys. Soc. Abstr. 9 (24): 172-173.

Anon. (1975). A summary of the ERMAC work session on nervous system and behavioral effects of nonionizing electromagnetic radiations. J. Microwave Power 10: 127-140.

Aristarkhov, V.M. et al. 1974. Changes in iron content under the influence of a permanent magnetic field of organs and tissues of intact animals and tumor-bearing animals. Akademiia Nauk SSR Izvestya Seriya Biol. 3: 429. (In Russian with English summary.)

Athenstaedt, H. 1970. Permanent longitudinal electric polarization and pyroelectric behavior of collagenous structures and nervous tissue in man and other vertebrates. Nature 228: 830.

Bankoske, J.W. and McKee, G. 1975. "Ecological Influences of Electric Fields." EPRI Project 129, Interim Report of May. Electric Power Research Inst., Palo Alto, Calif.

Barnes, H.C. 1972. Discussion of Paper 23.06 at CIGRE, 1972. CIGRE Proc. 1, Group 23. Reprinted in Special Publication #10, IEEE Power Engineering Society, Piscataway, N.J. (1975), G. Knickerbocker, ed.

Barron, C.I. and Dreher, J.J. 1964. Effects of electric fields and negative ion concentrations on test pilots. Aerosp. Med. 35: 20.

Barthold, L.O. et al. 1971b. "Electrostatic Effects of Overhead Transmission Lines. Part II. Methods of Calculation." IEEE

Transaction Paper 71 TP 645-PWR. Transmission and Distribution Committee of the IEEE Power Engineering Society.

Becker, J.F. *et al*. 1973. Orientation of chlorophyll *in vivo*. Studies with magnetic field oriented chlorella. *Biochem. Biophys. Res. Comm.* *51*: 597.

Becker, R.O. 1969. "The Effect of Mangetic Fields upon the Central Nervous System." IN: "Biological Effects of Magnetic Fields," Vol. 2. M. Barnothy, ed. Plenum Press, New York, p.-207

Becker, R.O. 1972a. Augmentation of regenerative healing in man. *Clin. Orthop.* *83*: 255.

Becker, R.O. 1972b. Electromagnetic forces and life processes. *MIT Tech. Rev.* *75* (December).

Becker, R.O. 1974c. The significance of bioelectric potentials. *Bioelectrochem. Bioenergetics* *1*: 187.

Becker, R.O., Chapin, S. and Sherry, R. 1974. Regeneration of the ventricular myocardium in amphibians. *Nature* *248*: 145.

Becker, R.O. and Murray, D.G. 1967. A method for producing cellular redifferentiation by means of very small electrical currents. *Trans. N.Y. Acad. Sci Series II* *29*: 606.

Beischer, D.E. 1974. "Literature Abstracts on Extremely Low Frequency (ELF) Electromagnetic Fields." The Neurosciences Research Program, Session on Brain Interactions with Weak Electric and Magnetic Fields (Nov. 10-12, 1974), 35 pp.

Beleslin, B.B. and Majić, V.V. 1974. The effects of pulsed magnetic field on the nerve cells of leeches. *Period. Biol.* *76*: 39.

Blumenthal, R., Changeux, J.P. and Lefever, R. 1970. Membrane excitability and dissipative instabilities. *J. Membr. Biol.* *2*: 351.

Bridges, J.E. *et al*. 1971. "Susceptibility of Cardiac Pacemakers to ELF Magnetic Fields." IITRI Project E-6185. Available from NTIS as AD 737237.

Brown, F.A. 1962b. Response of the planarian, *Dugesia*, and the protozoan, *Paramecium*, to very weak horizontal magnetic fields. *Biol. Bull.* *123*: 264.

Brown, F.A. and Chow, C.S. 1973. Interorganismic and environmental influences through extremely weak electromagnetic fields *Biol. Bull.* *144*: 437-461.

Cahill, L. 1972. Discussion of Paper 23.06 at CIGRE, 1972. CIGRE Proc. 1, Group 23. Reprinted in *Special Publication #10*,

IEEE Power Engineering Society, Piscataway, N.J. (1975), G. Knickerbocker, ed.

Caleca, V., Kolcio, and Shih. 1972. Discussion of Paper 23.06 at CIGRE, 1972. CIGRE Proc. 1, Group 23. Reprinted in Special Publication #10, IEEE Power Engineering Society, Piscataway, N.J. (1975), G. Knickerbocker, ed.

Chalazonitis, N., Chagneux, R., and Arvanitaki, A. 1970. Rotation des segments externes des photo recepteurs dans le champ magnetique constant. (Rotation of the outer segments of photo-receptors in a constant magnetic field.) C. R. Acad. Sci. Paris 271: 130.

Changeux, J-P. et al. 1967. On the cooperativity of biological membranes. Proc. Nat'l Acad. Sci. 57: 335.

Chen, C.S. and Pohl, H.A. 1974. Biologic dielectrophoresis: the behavior of lone cells in a non-uniform electric field. Ann. N.Y. Acad. Sci. 238: 176-185.

Chernysheva, M.A. 1975. "Effects of a Variable Magnetic Field of Industrial Frequency (50 Hz) on Metabolic Processes in the Organs of Rats." Gigiyena Truda, No. 11: 20-23 (in Russian). Translated in "Effects of Non-Ionizing Electromagnetic Radiation," and available from NTIS as JPRS L/5615 (10 Feb., 1976), pp. 33-37.

Chizenkova, R.A. 1975. "The Role of Different Brain Formations in EEG Reaction of Rabbits to a Constant Magnetic Field and to Ultrahigh- and Superhigh-Frequency Electromagnetic Fields." Translated in "Effects of Non-Ionizing Electromagnetic Radiation," and available from NTIS as JPRS 64532, pp. 10-22.

Cone, C.D. and Tongier, M. 1971. Control of somatic cell mitosis by stimulated changes in the transmembrane potential level. Oncogenesis 25: 168.

Cope, F.W. 1974. Enhancement by high electric fields of superconduction in organic and biological solids at room temperature and the role in nerve conduction. Physiol. Chem. Phys. 6: 405.

Dalziel, C.F. and Lee, W.R. 1968. "Re-evaluation of Lethal Electric Current." IEEE Industrial Trans. and Gen. Appl., Vol. IGA-4, Nos. 5 and 6. (Sept.-Oct. and Nov.-Dec.).

Duncan, R.I. and Macmillan, A.D. 1975. Embryonic chick tibia in steady electric fields. Nature 257: 162.

Einholf, C.E. and Carstensen, E.L. 1971. Low-frequency dielectric dispersion in suspensions of ion-exchange resins. J. Phys. Chem. 75: 1091.

Eisemann, B. von. 1975. "Untersuchungen über Langzeiteinwirkung Kleiner Wechselstrome 50 Hz auf den Menschen." Inaugural - Dissertation, Albert Ludwigs-Universität, Freiberg i. Br., Germany. (Also, Forschungsstelle fur Elektropathologie, 78 Freiburg I. Br., Im Gärtle 11, Germany.)

Electric Research Council. 1971. "Electric Utilities Industry Research and Development Goals through the Year 2000." Report to the R & D Goals Task Force to the Electric Research Council, ERC Publ. No. 1-71.

Fern, W.J. and Brabets, R.I. 1974. "Field Investigation of Ozone Adjacent to High Voltage Transmission Lines." IEEE Trans. Paper T-74 057-6, presented at IEEE Power Engineering Society Winter Meeting, New York, N.Y. (1974).

Frei, E.H. 1972. "Biomagnetics." IEEE Trans. on Magnetics MAG-8: 407-413.

Frydman, M., Levy, A., and Miller, S.E. 1972. "Oxidant Measurements in the Vicinity of Energized 765 kV Lines." IEEE Trans. Paper T-72 551-0, presented at IEEE Power Engineering Society Summer Meeting, San Francisco, Cal. (1972).

Frydman, M., and Shih, C.H. 1973. "Effects of the Environment on Oxidants Production in AC Corona." IEEE Trans. Paper T-73 407-4, presented at IEEE Power Engineering Society Summer Meeting and EHV/UHV Conf., Van Couver, B.C., Canada (1973).

Gardner, W.R., Harris, R.F. and Tanner, C.B. 1975. "Responses of Plants and Soil Microorganisms to Extremely Low Frequency Electric Fields." Univ. of Wisconsin- Madison, Final Report.

Gavalas-Medici, R. and Day-Magdaleno, S.R. 1976. Extremely low frequency, weak electric fields affect schedule-controlled behavior of monkeys. Nature 261: 256-259.

Giarola, A.J. and Krueger, W.F.1974. Continuous exposure of chic and rats to electromagnetic fields. IEEE Trans. on Micro. Th. & Tech. MTT 22: 432-437.

Goodman, E.M., Greenebaum, B. and Marron, M.T. 1976. Effects of extremely low frequency electromagnetic fields on Physarum polycephalum. Radiation Res. 66: 531-540.

Goodman, E.M., Marron, M.T., and Greenebaum, B. 1974. Long term effects of (ELF) electromagnetic fields on Physarum polycephalum. J. of Cell Biol. 63: 117.

Greenberg, B. 1973. Oxygen consumption in four species of invertebrates and vertebrates naturally exposed to Sanguine electromagnetic fields. Univ. of Ill.-Chicago Circle. Available from NTIS as ADA 001925.†

Greenberg, B. and Ash, N. 1973. "Impact of Extremely Low Frequency Electromagnetic Fields on Soil Arthropods." Ongoing studies of the Project Sanguine Test Facility, 1973, Univ. of Illinois-Chicago Circle. Available from NTIS as ADA 001099.†

Grissett, J.D. 1971. "Exposure of Squirrel Monkeys for Long Periods to Extremely Low Frequency Magnetic Fields: Central Nervous System Effects as Measured by Reaction Time." Naval Aerospace Medical Research Lab. NAMRL-1146. Available from NTIS as AD 735456.†

Grodsky, I. T. 1976. Neuronal membrane: a physical synthesis. *Math. Biosci.* *28*: 191-219.

Guillemin, R. *et al.* 1958. Measurement of free corticosteroids in rat plasma: physiological validation of a method. *Endicrinol.* *63*: 349.

Hainer, V. *et al.* 1972. Plasma corticosterone changes in starved rats: effect of thyroidectomy. *Horm. Metab. Res.* *4*: 314.

Halberg, F. 1959. Physiologic 24-hour periodicity: general and procedural considerations with reference to the adrenal cycle. *Z. Vitamin-Hormon. und Ferment Forschung* *10*: 225.

Halberg, F., Cutkomp, L., Nelson, W. and Sothern, R. 1975. "Circadian Rhythms in Plants, Insects and Mammals Exposed to ELF Magnetic and/or Electric Fields and Currents." Report, Minnesota Univ., Minneapolis Chronobiology Labs. AD A019-958.

Hauf, G. 1974a. Investigations on the action of technical energy fields on man. *Sonderdruck aus Elektrotechnische Zeitung* *26*, *H.* *12*: 318-320.

Hauf, G. 1974b. "Untersuchungen über Langzeiteinwirkung Kleiner WechWechselstrome 50 Hz auf den Menschen." Inaugural - Dissertation, Albert Ludwigs-Universität, Freiberg i. Br., Germany. (Also, Forschungsstelle fur Elektropathologie, 78 Freiberg I. Br., Im Gärtle 11, Germany.)

Hauf, R. 1976. Einfluss elektromagnetischer Felder auf den Menschen. *Sonderdruck aus etz-b Bd.* *28*, *H.* *6/7*: 181-183.

Hauf, R. and Wiesenger, J. 1973. Biological effects of technical electric and electromagnetic VLF fields. *Int. J. Biometeor.* *17*: 213-215.

Heibel, J.T. *et al.* 1972. "Measurements of Effluents Due to EHV Transmission Line Corona, I." Field tests presented at IEEE Canadian Communications and EHV Conference, Montreal, Canada.

Horwitz, B.A. and Horwitz, L.P. 1973. Electronic distribution coding in cellular differentiation. *J. Theor. Biol.* *42*: 169.

Hubbard, H.I. 1974. "The Peripheral Nervous System." Plenum Press, New York.

Irwin, D.D., Rush, S., Evering, R., Lepeschkin, E., Montgomery, D., and Weggel, R.J. 1970. Stimulation of cardiac muscle by a time-varying magnetic field. IEEE Trans. on Magnetics MAG-6: 321-322.

Kacmarek, L.K. and Adey, W.R. 1973. The efflux of $^{45}Ca^{2+}$ and (^{3}H)γ-aminobutyric acid from cat cerebral cortex. Brain Res. 63: 331.

Kemeny, G. and Rosenberg, B. 1970. Theory of the pre-exponential factor in organic semi-conductors. J. Chem. Phys. 52: 4151.

Klimovskaya, L.D. and Smirnova, N.P. 1975. Some autonomic reactions in rabbits exposed to a permanent magnetic field. Space Biology and Aerospace Medicine 9: 18-22. (JPRS 65301).

Kolin, A. 1968. Magnetic fields in biology. Physics Today 21: 39

Kolin, A., Brill, N.Q. and Broberg, P.J. 1959. Stimulation of irritable tissues by means of an alternating magnetic field. Proc. Soc. Exp. Biol. and Med. 102: 251-252.

Kolodub, F.A. and Yevtushenko, G.I. 1972. Gigiena Truda i Professional'nyye Zabolevaniya."Biochemical aspects of the Biological Effect of a Low-Frequency Pulsed Electromagnetic field." Available from NTIS as JPRS 56583.

Kolta, P. 1973. Strong and permanent interaction between peripheral nerve and a constant inhomogeneous magnetic field. Acta Physiologica Akademiae Scientiarum Hungaricae 43: 89.

Kotaka, S. and Krueger, A.P. 1967. Studies on the air-ion-induced growth increase in higher plants. Adv. Front Plant Sci. 20: 115.

Kovalenko, O.A. et al. 1974. Bioritmy: I. Tsirkadnye ritmy kontsentratsii paramagnitykh tsentrov pecheni mysheiv norme (Biorhythms: I. Circadian rhythms of a concentration of paramagnetic sensors in the normal mouse liver). Biofizika 19: 488. (In Russian with English summary). Abstract in Biological Abstracts 59: 2598 (#23318)(1974).

Krasov, V.M. et al. 1967. "The Nature of the Effect of a Magnetic Field on Nutrient Medium Composition and Toxin Production by Cl Perfringens, Type D." Konferentsiya Biokhimikov Respublik Sredney Azii I Kazakhstana. Available from NTIS as "Translations on Biological Effects of Magnetic Fields," JPRS 62865.

Krebs, J.R. (rapporteur) 1975. How birds reach home. Report on 14th Internat'l Ethological Conf. Nature 257: 338.

Krueger, A.P. and Reed, E.J. 1975. "A Study of the Biological Effects of Certain Electromangetic Fields." Report from Univ. of California--Berkeley.

Krueger, A.P. and Smith, R.F. 1960. The biological mechanisms of air ion action, I. 5-hyroxy tryptamine as the endogenous mediator of positive ion effects on the mammalian trachea. J. Gen. Physiol. 43: 533.

Lang, S. 1970. "Investigation on the Behavioral Physiological and Metabolic Physiological Effects of the Faraday Screening of Artificial Atmospheric Electrical Direct and Alternating Fields on White Mice (Mus musculus)." Dissertation, Universität des Saarlandes. (In German).

Lang, S. 1972. The metabolic and physiological effects of Faraday screening and of an artificial air electric field of a frequency of 10 Hz on white mice. Arch. Met. Geoph. Biokl. (ser B) 20: 109-122. (In German).

Lawler, J.C., Davis, M.J. and Griffith, E.C. 1959. Electrical characteristics of the skin. J. Invest. Dermatol. 34: 301.

Liberman, E.A. 1958. Possible ways of detecting electron conduction in nervous elements. Biofizika 3: 743.

Liepa, M.E. and Slutskii, L.I. 1974. The effect of a constant magnetic field on the healing process of wounds. (In Russian). Latv. Psr. Zinat. Akad. Vestis. 5: 140. Abstract in Biological Abstracts 59: 3130 (1975).

Ludwig, H.W. 1971. The effect of electromagnetic extremely low frequency alternating fields on higher organisms. Biomedizinische Tecknik 16: 67-72. (In German).

Ludwig, H.W. 1974. "Electric and Magnetic Field Strengths in the Open and in Shielded Rooms in the ULF-LF Zone." Translated by I. Hauk IN: "ELF and VLF Electromagnetic Field Effects." M. Persinger, ed. Plenum Press, New York, p. 35.

Maass, J.A. and Asa, M.M. 1970. Contactless nerve stimulation and signal detection by inductive transducer. IEEE Trans. on Magnetics MAG-7: 322-326.

Mamontov, S.G. and Ivanova, L.N. 1971. Effect of a low-frequency electric field on cell division in mouse tissues. Bull. Exp. Biol. and Med. 71: 192-193.

Mantell, B. 1975. "Untersuchungen über die Wirkung eines magnetischen Wechselfeldes 50 Hz auf den Menschen." Inaugural-Dissertation, Albert Ludwigs-Universität, Freiberg i. Br., Germany. (Also, Forschungsstelle fur Elektropathologie, 78 Freiburg I. Br., Im Gärtle 11, Germany.)

Marino, A.A. 1974. Prepared testimony before the State of New York Public Service Commission, Case 26559.

Marino, A.A., Berger, T.J., Becker, R.O., and Hart, F.X. 1974c. Electrostatic field induced changes in mouse serum proteins. Experientia 30: 1274-1275.

Marx, J.L. 1975. Air pollution: effects on plants. Science 187: 731.

May, L., Baumgartner, C. and Cuesta, E.R. 1973. The effect of electric fields on brain lipid-cholesterol films. J. Membr. Biol. 14: 63.

May, J.L., Kamble, A.B. and Acosta, I.P. 1970. The effect of electric fields on brain cephalin and lecithin films. J. Membr. Biol. 2: 192.

McCleave, J.D., Albert, E.H. and Richardson, N.E. 1974. "Perception and Effects on Locomotor Activity in American Eels and Atlantic Salmon of Extremely Low Frequency Electric and Magnetic Fields." Univ. of Maine, Final Report. Available from NTIS as AD 778021.†

McGinness, J., Corry, P. and Procter, P. 1974. Amorphous semiconductor switching in melanins. Science 183: 853.

Meilinger, H. 1975. "Effect of an Electric Stress on Microorganisms." IN: "Response of Plants and Soil Microorganisms to ELF Fields." W. Gardner, ed. Report for Project Sanguine. Dept. of Soil Science, Univ. of Wisconsin, Madison, Wis.

Miller, C.J. 1967. The Measurement of Electric Fields in Live Line Working. IEEE Trans. Pow. App. and Sys. PAS-86, No. 4.

Mittler, S. 1971. Failure of magnetism to influence production of x-ray induced, sex-linked recessive lethals. Mutat. Res. 13: 287.

Molnar, K. 1972/73. Sub-biological aspects of ageing and the concept of biological cathode protection. Mech. Ageing Dev. 1: 319.

Nahas, G.G., Boccalon, H., Berryer, P. and Wagner, B. 1975. Effects in rodents of a 1-month exposure to magnetic fields (200-1200 gauss). Aviation, Space and Environ. Med. 46: 1161-1163

Neumann, E. and Katchalsky, A. 1971. Hysteretic conformational changes in biopolymers induced by high electric fields -- model for a physical record of biological memory. Proc. 1st Eur. Biophys. Cong. (Baden/Vienna) 6: 91.

Ochs, A.L. and Burton, R.M. 1974. Electrical response to vibration of a lipid bilayer membrane. Biophys. J. 14: 473.

Pereira, M.R. et al. 1967. Cellular respiration in intermittent magnetic fields. Prod. Soc. Exp. Biol. Med. 124: 573.

Persinger, M.A., Lafreniere, P.F. and Ossenkopp, K-P. 1974. "Behavioral, Physiological and Histological Changes in Rats Exposed during Various Developmental Stages to ELF Magnetic Fields." IN: "ELF and VLF Electromagnetic Field Effects." M. Persinger, ed. Plenum Press, New York, p. 177.

Pilla, A.S. 1974. Electrochemical information transfer at living cell membranes. Ann. N.Y. Acad. Sci. 238: 149-170.

Piruzyan, L.A. et al. 1970. "The Mechanism of the Biological Effect of Constant Magnetic Fields." Izvestiya Akademii Nauk SSSR, Seriya Biologicheskaya, No. 4. Available from NTIS as "Biological Effects of Constant Magnetic Fields," JPRS 57927. (1973).

Piruzyan, L.E. et al. 1972. "The Influence of a Constant Magnetic Field on the Concentration of Free Radicals in the Organs and Tissues of Mice." Izvestiya Akademii Nauk SSSR, Seriya Biologicheskaya, No. 1. Available from NTIS as "Biological Effects of Constant Magnetic Fields," JPRS 27927 (1973).

Roach, J.F., Chartier, V.L. and Dietrich, F.M. 1973. "Experimental Oxidant Production Rates for EHV Transmission Lines and Theoretical Estimates of Ozone Concentrations near Operating Lines." IEEE Trans. Paper T-73 414-0, presented at IEEE Power Engineering Society Summer Meeting and EHV-UHV Conf., Vancouver, Canada.

Rupilius, J.P. 1976. "Untersuchungen über die Wirkung eines elektrischen und magnetischen 50 Hz-Wechselfeldes auf den Menschen." Inaugural - Dissertation, Albert-Ludwigs-Universität, Freiburg i. Br., Germany. (Also, Forschungsstelle fur Elektropathologie, 78 Freiburg, I. Br., Im Gärtle 11, Germany.)

Sandifer, J.R. and Buck, R.P. 1975. An algorithm for simulation of transient and alternating current, electrical properties of conducting membranes, junctions, and one-dimensional, finite galvanic cells. J. Phys. Chem. 79: 384.

Sandler, S.S., Smith, G.S. and Albert, E.N. 1975. Electromagnetic field effects in nerve tissue. Aviation, Space and Environ. Med. 46: 1414-1417.

Scherer, H.N., Ware, B.J. and Shih, C.H. 1972. "Gaseous Effluents due to EHV Transmission Line Corona." Trans. Paper T 72 550-2, presented at the IEEE Power Engineering Society Summer Meeting, San Francisco, California.

Schmitt, O.H. 1973. "Living in Magnetic Fields." Biophysics Group Electric. Eng., Univ. of Minnesota, Minneapolis, Minn. 55455.

Scott, B.I.H. and Gulline, H.F. 1975. Membrane changes in a circadian system. Nature 254: 69.

Sebo, S.A. et al. 1972. "Measurements of Effluents due to EHV Transmission Line - II. Laboratory Tests." Proc. IEEE Canadian Communications and EHV Conf., Montreal, Canada.

Shul'pekov, A.A. 1971. "Peculiarities of Methods and Methodology of Magnetobiological Experiments." IN: "Influence of Magnetic Fields on Biological Objects." Y. Kholodov, ed. Available from NTIS as JPRS 63038 (1974).

Singer, H. 1972. Discussion of Paper 23.06 at CIGRE, 1972. CIGRE Proc. 1, Group 23. Reprinted in Special Publication #10, IEEE Power Engineering Society, Piscataway, N.J. (1975), G. Knickerbocker, ed.

Singer, M. 1952. The influence of the nerve in regeneration of the amphibian extremity. Quart. Rev. Biol. 27: 169.

Singer, S.J. and Nicolson, G.L. 1972. The fluid mosaic model of the structure of cell membranes. Science 175: 720.

Spiegel, R.J. and Joines, W.T. 1973. A semi-classical theory for nerve excitation by a low intensity electromagnetic field. Bull. Math. Biol. 35: 591.

Spittka, O., Taege, M. and Tembrock, G. 1969. Experimental investigations on the operant drinking behavior of rats in the 50 Hz high tension alternating field. Biol. Zbl. 88: 273-282.

Stamenović, B.A. and Majić, V.V. 1975. Heart function in magnetic field. Period. Biol. 77: 57-61.

Strelow, R. 1975. Extremely high voltage transmission lines - health and environmental effects: request for submission of data. Fed. Reg. 40 (#53): 12312 (Mar. 18, 1975).

Tomaselli, V.P. 1974. Electrical properties of hydrated collagen II. Semi-conductor properties. Biopolymers 13: 2323.

Tredgold, R.H. and Sproule, R.C. 1973. Dielectric hysterisis in proteins. Biochimica et Biophysica Acta 317: 569.

Treager, R.T. 1965. Interpretation of skin impedence measurements. Nature 205: 600.

Tronchetti, F., Marescotti, V. and Saba, P. 1967. "Control Mechanism of Circadian Rhythm of Corticosterone in Adrenal and Peripheral Blood in the Rat." Proc. 2nd Internat'l Cong. on Hormonal Steroids, L. Martini, ed.

Tsoneva, M.T. *et al*. 1975. The effect of magnetic field on chromosome set and cell division. *Genetika* *11*: 153.

Watson, J., DeHaas, W.G., and Hauser, S.S. 1975. Effect of electric fields on growth rate of embryonic chick tibia *in vitro*. *Nature*: *254*: 331-332.

Zervins, A. 1973. Chick embryo development in 26-Hz electromagnetic field. *Am*. *Ind*. *Hyg*. *Assn*., p. 120 (Mar.).